Our future inheritance: choice or chance?

A STUDY BY A BRITISH ASSOCIATION
WORKING PARTY

BY
ALUN JONES
AND
WALTER F. BODMER

OXFORD UNIVERSITY PRESS
1974

Oxford University Press, Ely House, London W. 1

GLASGOW NEW YORK TORONTO MELBOURNE WELLINGTON
CAPE TOWN IBADAN NAIROBI DAR ES SALAAM LUSAKA ADDIS ABABA
DELHI BOMBAY CALCUTTA MADRAS KARACHI LAHORE DACCA
KUALA LUMPUR SINGAPORE HONG KONG TOKYO

CASEBOUND ISBN 0 19 857384 7
PAPERBACK ISBN 0 19 857390 1

© OXFORD UNIVERSITY PRESS 1974

All rights reserved. No part of this publication may be reproduced, stored in a retrieval system, or transmitted, in any form or by any means, electronic, mechanical, photocopying, recording or otherwise, without the prior permission of Oxford University Press.

F2177 £1·25 2·25

PRINTED IN GREAT BRITAIN BY
RICHARD CLAY (THE CHAUCER PRESS), LTD., BUNGAY, SUFFOLK

Preface

THE seeds for this book were sown at the end of 1971 when there were many discussions as to how the British Association for the Advancement of Science should meet the growing challenges of the 1970s. Until then the Association had largely concentrated its efforts on its annual meeting as a means of making the public aware of the developments and consequences of science. The annual meeting did, and still does, fulfil a most useful purpose, but it was felt that more was needed in some areas of science which were developing particularly quickly and had important potential consequences to society.

It was Section X of the Association, the section concerned with general topics, that set up a Committee on Science and Public Affairs, which in its turn set up a working part to study, as it was then put to prospective members, 'the scientific, social, ethical and legal implications of recent advances in genetics and biology', with the ultimate objective of publishing a book on these implications. One of us (W.F.B.) was the chairman and the other author (A.J.) the secretary. During the next two years, several meetings of the working party were held and an impressive number of experts in several fields contributed either by preparing working papers as bases for discussion or by attending the meetings.

From the beginning of 1972 the working party was left free by the British Association to follow its own ideas, although the association's secretary, Dr. J. A. V. Willis, was a regular member until his retirement in September 1973. The Committee on Science and Public Affairs, like all good mothers, also encouraged the working party to develop under its own momentum. In November 1973, a few months before this book went to press, the committee ceased to exist, its appointed tasks having been subsumed by the council of the Association under its new secretary, Dr. Magnus Pyke. In 1974, the working party is facing other challenges as the Committee for Social Concern and Biological Advances, with its continued existence assured through the generosity of the Leverhulme Trust. The members had built up such a sense of community that it was unanimously felt that there was a need in Britain for such a multidisciplinary group, if for no other reason than to make lawyers, politicians, churchmen, and scientists aware of each other's problems.

The enthusiasm of all who attended the eight meetings held between the beginning of 1972 and the end of 1973 cannot easily be captured and transmitted in a book. Any failure here is solely the responsibility of the authors and is in no way a reflection on the many people who contributed their help. The content and form of this book might not meet with the full approval of everyone who has been associated with its gestation. But those members of the working party who have regularly attended the meetings and have been significant contributors, although they are not all listed as co-authors (see the lists below), are glad to sanction its publication.

It will inevitably be asked for whom this book is intended. The answer is that the working party feels, and the authors hope, that it will be read by those who are interested in the way in which modern biology and genetics have developed in recent years but especially to those who are also interested in the way in which it will develop in the coming years. First and foremost, the book is not intended to be exclusively for biological scientists although it is hoped that these scientists will feel inclined to acquaint themselves with its contents. We also hope that this book will be useful in educating the public in recent developments in the biological sciences. The lay public, whatever such a nebulous term may imply, will gain a great deal from discovering what is going on at the so-called frontiers of research and from sharing with those directly concerned the dilemmas of applying the results of these researches. To be forewarned is to be prepared and if at some time in the future any of the topics raised in this book become emotional issues, receiving a great deal of publicity, then the background to the work is to be found here.

Each of the fact-finding meetings of the working party has resulted in a chapter of this book. As at the meetings, the scientific issues are first presented and these are followed by a discussion of the problems of their application.

Preface

We make no pretence that the coverage of topics has been comprehensive; this book chiefly reflects the particular expertise and interest of the members of the working party. There is no doubt that other topics, for example the effect of behavioural control drugs, also raise important issues; but these will have to be dealt with at some other time.

The regular members of the working party who attended the majority of the meetings were: Professor W. F. Bodmer, Department of Genetics, University of Oxford; Professor C. R. Austin, Physiological Laboratory, University of Cambridge; Professor A. E. Boyo, Department of Medicine, University of Lagos (who spent 1972 on sabbatical leave at King's College, Cambridge); Professor G. R. Dunstan, Faculty of Theology, King's College, University of London; Professor J. H. Edwards, Department of Human Genetics, University of Birmingham; Dr. R. G. Edwards, Physiological Laboratory, University of Cambridge; Dr. Alun Jones, *Nature*; Dr. Anne McLaren, Department of Genetics, University of Edinburgh; Mr. John Maddox, *Nature* and later Maddox Editorial; Dr. David Owen, Member of Parliament; Mrs. Shirley Williams, Member of Parliament; and Dr. J. A. V. Willis, British Association for the Advancement of Science.

Many others attended one or two meetings. These were: Professor G. H. Beale, Institute of Animal Genetics, University of Edinburgh; Professor R. Y. Calne, Addenbrooke's Hospital, University of Cambridge; Dr. C. C. Carter, M.R.C. Clinical Genetics Unit, London; Dr. M. d'A. Crawfurd, Department of Genetics, University of Leeds; Mr. D. Deadman, Press Council; Professor M. A. Ferguson-Smith, Department of Genetics, University of Glasgow; Sir George Godber, Department of Health and Social Security; Dr. Christopher Graham, Department of Zoology, University of Oxford; Sir John Gray, Medical Research Council; Mr. J. Selwyn Gummer, Member of Parliament; Dr. Margaret Jackson, medical practitioner, Devon; Professor T. E. James, at one time of the Faculty of Laws, King's College, University of London; Dr. George Komrower, consultant paediatrician to the United Manchester Hospitals and Manchester Regional Hospital Board; Professor Richard Lewontin, Department of Genetics, University of Chicago and now of Harvard University; Professor Sir Alan Parkes, The Galton Foundation, London; Dr. Magnus Pyke, British Association for the Advancement of Science; Professor Ralph Riley, Plant Breeding Institute, Cambridge; Professor Steven Rose, Department of Biology, Open University; Dr. Eliot Slater, Institute of Psychiatry, London; Mr. C. P. Steptoe, Consultant Surgeon, Oldham and District General Hospital; Professor Peter Townsend, Department of Sociology, University of Essex; Mr. P. D. Trevor-Roper, Consultant Surgeon, Westminster Hospital; and Professor Sir Michael Woodruff, Department of Surgery, University of Edinburgh Medical School.

The following also prepared working papers which formed the basis for discussion; Professor Austin, Professor Beale, Professor Calne, Dr. Crawfurd, Mr. Deadman, Professor Dunstan, Professor J. H. Edwards, Professor Ferguson-Smith, Sir George Godber, Dr. Margaret Jackson, Professor James, Dr. Komrower, Professor Riley, Dr. Slater, Mr. Steptoe, and Mr. Trevor-Roper.

It should be made clear that this book is not an unconnected series of the working papers joined together by means of a distillation of the discussion at the meetings. The working papers indeed have formed the basis for the book and the discussions have placed the various written contributions in their proper context. The book is the work of the authors, who have tried to maintain a uniform style and standard of presentation. The regular members of the working party saw and read the manuscript and suggested modifications which, in general, were adopted.

We have made few attempts to draw conclusions from the facts which have been accumulated. Rather, questions are raised and anomalies pointed out. Possible dangers are flagged and, we hope, some myths are destroyed. It is not our intention to be controversial but only to point out where controversy does arise.

We are grateful to many people who at some time in the past two years have enabled the work on this book to progress. A.J. is particularly indebted to Mr. John Maddox for the opportunity to participate in the study and for the time to do some of the writing. We are also grateful to Miss Fiona Selkirk and Miss Cara Bulman for typing the manuscript. Last but not least, the book would not have been possible without the continuing help of the members of the working party and all others who contributed in various ways to the overwhelming success of every meeting.

We should like to thank the following who have given permission for the reproduction of illustrations and have lent originals:

Dr. A. C. Allison, M.R.C. Clinical Research Centre, Harrow. (Pl.

3 (*b*, *c*)); Dr. M. Bobrow, University of Oxford; Dr. R. G. Edwards, Physiological Laboratory, University of Cambridge (Pls. 1 and 2, from *American Journal of Obstetrics and Gynecology*, **117**(5), 587–601); Professor H. Harris, Department of Pathology, University of Oxford (pl. 4 (*b*), from *Nucleus and cytoplasm*, 2nd edition, Clarendon Press, 1970); Dr. G. W. Komrower, Royal Manchester Children's Hospital (Pl. 4 (*a*)).

Fig. 4 is adapted from L. Cavalli-Sforza: *Elements of human genetics* (Addison-Wesley); Fig. 4.5 is adapted from Dancis, 'The Prenatal Detection of Hereditary Defects' in McKusick and Claiborne: (ed). *Medical Genetics* (Hospital Practice Publishing Co.); Fig. 5.1 is based on data from the Eleventh Report of the Human Transplant Registry, published in the *Journal of the American Medical Association*, **226** (10), 1197–204 (1973); Fig. 6.1 is adapted from J. D. Bernal and Ann Synge: *The origin of life* (Oxford Biology Reader 13).

We are also grateful to Professor A. Etzioni and the editor of *Science* for permission to use an extract from Science, **161**, 1107-1112 (1968) © 1968 by the American Assoc. for the Advancement of Science.

Oxford   A.J.
January 1974   W.F.B.

Contents

List of Plates	xiii
1. Biomedical advances: a mixed blessing?	1
2. Artificial insemination by donor and the treatment of infertility	12
3. Artificial fertilization	30
4. Genetic screening and selective abortion	45
5. Organ transplantation	83
6. Genetic engineering and cloning	106
7. Social concern and biological advances	116
Suggestions for further reading	134
Index	137

List of Plates

The egg and sperm and *in vitro* fertilization

1 (a) The unfertilized egg
1 (b) Sperm about to penetrate the egg

2 (a) The four—cell stage-two cleavages after fertilization
2 (b) The human embryo 5 to 6 days after fertilization

between pp. 66–7

3 (a) The chromosome complement of a normal human male

Red blood cells of an individual with the sickle-cell trait:

3 (b) Red cells in their normal disc-shaped state in the presence of oxygen
3 (c) The same cells when oxygen has been removed

4 (a) Screening for phenylketonuria (PKU)
4 (b) A cell formed by fusion of human and mouse cells

between pp. 82–3

1 Biomedical advances: a mixed blessing?

THERE are many and varied consequences of scientific research but few would deny that the one inescapable result of research is that it increases knowledge and that the application of this knowledge in general can improve the quality of life. This axiom was of course true even long before science as it is known today was ever practised. There is also another inescapable result of research: that scientific discoveries will not always be applied to what most people would consider to be desirable ends.

It would be a utopian society if all applications of science produced nothing but good and it is foolish to try to disguise the fact that there can be unsavoury or unsatisfactory applications of many scientific advances. Even the application of such advances in what would be considered advantageous ways could have other, unsatisfactory, effects on society. In fact, Sir Isaiah Berlin's warning in his essay on historical inevitability that there is danger in assuming that 'men will know more and therefore be wiser and better and happier' is particularly relevant.

This conflict between the savoury and the unsavoury is the essential nature of the application of science. What is of importance is that this nature is both recognized and understood and that discussion of the issues is not suppressed, either through ignorance or for any devious motive.

There are some clear examples of how the application of scientific discoveries has led to far-from-utopian solutions to various problems. For example, the harnessing of nuclear power in the 1940s, provided in principle, an abundant supply of energy, but the use of this knowledge to create nuclear explosives has posed one of the greatest challenges to society. As a further example, the use of D.D.T. to exterminate mosquitoes and thus to counter the effects of malaria in Africa and elsewhere would be considered a most desirable development of science, but the realization that D.D.T. accumulates in the environment has led to demands for banning its use.

The potential advantages and disadvantages of scientific advances cannot always be predicted at the time of the discovery. This, however, is no reason why the various possibilities should not be discussed, and it is important that the discussion is not restricted to a closed community of scientists. While scientists are palpably the best (and the only) people to carry out the research, they are not necessarily the best people to discuss its implications. In fact, the thinking and type of decision necessary before a scientific development is applied in society is very different from the type of thinking to which a scientist is accustomed.

The way in which a scientist thinks and operates is quite different from the way in which a doctor or a lawyer, for example, practises his profession. A scientist will, in general, attack a problem when he believes a solution to be possible with known methodology; or at the very most he will proceed far enough with a problem to see whether it is worth going further. According to Sir Peter Medawar, one of the 1960 Nobel Laureates in physiology and medicine, 'science is the art of the soluble'.

Lawyers and doctors have to approach their problems differently. When they are faced with a decision on whether or not to prosecute, how to sentence, how to treat a patient, whether to operate, where to cut, and so on, they have to make a decision however inadequate the bases for proceeding may be.

This difference in approach is exemplified in the way the different groups—scientists, doctors, and lawyers—would define the start of life. A biologist might say that human life started at the moment of fertilization when the sperm and the ovum merged. A doctor, on the other hand, would say that it is not possible to say that a woman is pregnant until several days later, when the embryo attaches itself to the wall of the uterus; so they could argue for this time as the start to life. Both these interpretations are impracticable for lawyers, who have to make decisions which will work in society, and so they take birth to be the starting-point of life.

Almost any field of science could be taken and studied in detail to reveal possible future unsatisfactory applications of research. In recent years, however, the developments of modern biology and genetics in particular have aroused many fears. This no doubt has been helped along by such journalistic phrases as 'genetic engineering' and 'test-tube babies'. There has unfortunately been little informed discussion of the implications of this research—both advantageous and otherwise—and there has been little attempt to put forward a balanced picture of the probable advantages and possible disadvantages of research now in progress.

Although concern for such biomedical advances is not new—history is strewn with examples of scientists and doctors being harassed because of their work—it is probable that such concern today is heightened because of two factors. First, communication is easier through radio, television, daily newspapers, and magazines. Second, there is an ever-increasing rate of scientific discovery and technological advance. The result is that concepts are put forward before even the basic research is completed, so that inevitably there has to be some speculation as to how the work will be applied. This, of course, is no bad thing. It is important that people should be kept appraised of what the scientists are doing in their laboratories, if for no other reason than that much of the money being spent comes from public funds. This brings out the need to ensure that the standards of communication are high so that the reports carried by the media are equitable.

The great majority of doctors and scientists involved in research and in the further development of scientific results are more than keen that their work and its implications should get a public airing. Nothing is more exasperating to a scientist than to see his work put into a context in an article for the lay public that, if not completely misrepresenting the aims, at least puts the emphasis on aspects that are not feasible, or practicable, and do not give the chief objective due attention.

In the chapters that follow the current state of scientific knowledge in some key fields of medical and genetic research is presented in order to place articles, statements, or comments on these topics in proper perspective. By way of introduction, the salient points are now summarized.

Artificial insemination

Aiding couples who for one reason or other are sterile to have children is an emotive subject. Artificial insemination by the sperm of a donor has in the past 30 years made headline news more than once, and in spite of all the public discussion and the high-powered commissions which have looked into A.I.D. there is still no public policy on the practice. The legal tangle which the practice of A.I.D. leads to is expounded in Chapter 2.

Recent developments in the freezing of sperm will surely lead to more and more inseminations in the future. No restrictions are placed on practitioners of A.I.D. and there is no code, either legal or otherwise, to which practitioners must adhere. Artificial insemination by the husband's sperm will also increase with the techniques now available for freezing and storing sperm. These provide a way in which subfertile husbands can accumulate and store their semen for insemination at the most apposite time in the wife's menstrual cycle.

There may be many side-effects of freezing and storing sperm. In the United States the chief use of frozen sperm banks, as they are called, is to provide a form of insurance for husbands about to undergo vasectomies. Another issue is to what extent, if any, should prospective customers of sperm banks be allowed to choose the sperm they want for artificial insemination. Or, to turn the question round another way, what qualities are advisable or necessary in sperm donors? In one of the many articles which have been written about the implication of frozen semen the possibility was raised of women queuing for the sperm of a latter-day Einstein. But behaviour, intelligence, and other characteristics are in fact determined by complex interactions between many genetic and other factors, such as the environment in which the child is brought up. It is therefore not clear that a *curriculum vitae* of the sperm donor will be of any help to a wife in making her choice.

Artificial fertilization

In many marriages the wife rather than the husband is the cause of the couple's infertility. Work is now going on in laboratories in various parts of the world to cure one particular form of the wife's infertility: that in which she has at least one healthy ovary and a healthy uterus, but the Fallopian tubes are either diseased or incapable of channelling the ova, or eggs, between the organs. This work has caused many a furore. In essence, the technique used, *in vitro* fertilization, consists of removing the ova from the ovaries by a minor operation, using a hypodermic needle inserted into the abdomen, and then fertilizing these in a dish with the husband's semen. This procedure has been accomplished successfully and 7-day-old fertilized eggs have been grown outside the body (see Pls. 1 and 2). Up to the end of 1973, however, there had been no reported success in transferring the young embryo back into the womb for it to develop and grow naturally, a procedure which has been made to work in some animals. The chief problem is that the hormone balance of a woman who is a few days into pregnancy has in some way to be faithfully reproduced before the embryo can attach itself to the wall of the uterus. As yet this hormone balance is not sufficiently well understood.

The fears which have been expressed about this work, apart from the fact that potential embryos a few days old 'have been washed down the sink', are based on the possibility—even probability—that the technique, once developed, will be abused. Disposing of the young embryos to those who consider life absolute from the moment of fertilization is no little matter, but the use of a coil as an interuterine contraceptive can be argued to have the same effect in that it does not prevent fertilization but it does prevent the embryo attaching itself to the wall of the uterus.

The possible abuse of this technique once it is developed has had much more publicity than the infertility aspect. The effects of a variety of permutations and combinations of sperm and ova have been widely written about, and the possibility has been aired of women providing foster wombs for children of mothers who are unable, or do not want, to spend 9 months carrying a child. While there should be no attempt to shy away from discussing these possible applications of this work, it should be considered in the context of its essential aims and the chief potential beneficiaries.

Genetic diseases

Illnesses caused by genetic factors are becoming more and more matters of public anxiety as 'conventional' illnesses become infrequent or even disappear altogether. Cholera and smallpox are now rare events indeed and outbreaks, when they do occur, make headline news. But medical science and public health organizations have developed to such a stage, at least in the developed world, that the outbreaks are soon brought under control and few, if any, lives are lost. Just as dramatic is the decline in tuberculosis and poliomyelitis since the mid-1950s and measles since the early 1960s. The result is that the relative proportion of people suffering from overt genetic illnesses is increasing and in recent years attention has gradually focused on these ailments.

The World Health Organization, in a report issued in 1972, stated that for one children's hospital in Montreal 11 per cent of all admissions were suffering from genetic diseases. There is every reason to suppose that these figures apply universally in the developed countries, and in all probability the relative proportion of children suffering from genetic diseases will increase with time as fewer and fewer people suffer from conventional illnesses in early life.

In contrast to infectious illnesses such as measles and poliomyelitis, the presence of inherited diseases can, in principle, be detected at a very early stage of development. To say that it can be done in principle means that the defect is present from conception, but for most diseases there is no way of detecting it until the overt symptoms appear. This may be when the child is several years old. For example, Huntington's chorea, a genetic disease causing severe nerve degeneration, does not show any outward effects until the sufferer is well into middle age. This is only one example of how genetic diseases differ from conventional illnesses. The common diseases of childhood seem to occur randomly and there is no way of telling before the illness strikes which child is going to be affected. (It is possible that there may be genetic factors which alter a person's susceptiblity to disease. Such factors, however, have not so far been identified for any of the common infectious diseases of childhood.) Mass inoculations and vaccinations have ensured that nearly all children are resistant to the common illnesses of childhood. The programme has worked well: these common diseases have become rare, and much suffering has been averted. Whether the decrease can be attributed solely to the effects of vaccinations and inoculations is a moot point. Diphtheria, for example, was declining quickly in the developed countries before immunization was introduced. Vaccination certainly had some effect but improvements in overall hygiene and nutrition also contributed substantially.

There is little that can be done to stop most of the genetically determined diseases from manifesting themselves once the child is conceived. Ideally, prevention should occur before conception. If this is not possible, diagnosis at an early stage in pregnancy must be undertaken; then if the foetus is shown to be affected the only recourse is to perform an abortion. This immediately creates many difficulties, for abortion is understandably a highly emotive topic. A question is asked whether any life is better than no life. A few genetic diseases are amenable to treatment by diet, transfusion, or drugs, but what is life like for the sufferers and their families? The conditions for each disease are different, and the prospects for treatment for the most common diseases are enlarged upon in Chapter 4 in order that these questions can be given full consideration.

Organ transplantation

A reluctance to shake off traditions has perhaps resulted in organ transplant operations receiving such mediocre publicity and support in recent years. Kidney transplantation is now a well-proved technique which gives the recipient the opportunity to live a near normal life. There is the capacity in British hospitals to carry out many more transplants than are undertaken at present but there is a severe shortage of kidneys. In the early days of kidney transplants the kidneys were obtained from very near relatives but most kidneys are now being obtained from cadavers. This is in part because of the success in matching or 'tissue typing' of kidneys and in part because of improvement in surgical techniques and the subsequent medical management of the patients. Although 7500 or so people are killed on British roads every year, very few of these kidneys become available for transplantation, where they could help to save lives. The law demands that all efforts must be made to obtain the permission of the deceased's next of kin before removing the kidneys. Allowance is made for the doctor to go ahead and remove the kidneys provided all reasonable efforts have been made to contact the relatives, but it is not surprising that doctors are loath to take this last option.

In essence, there is a great deal of ignorance about kidney transplants. Their success is not generally known, and in spite of attempts by the Department of Health and Social Security to publicize the demand for kidneys the increase in donors has not matched the demand. Other transplants, such as those of corneas and hearts, suffer from the same kind of public apathy. In Scandinavia, by contrast, kidneys can be taken from dead patients without consent and there is no lack of organs for transplantation. Will this ever happen in Britain?

'Genetic engineering'

On a futuristic note, Chapter 6 of this book deals with possible means of alleviating genetic diseases by 'genetic engineering'. Genetic engineering can be divided into two categories. First there is the selection and manipulation of sperm and ova to determine the genetic make-up of future offspring. Second, there is the manipulation of body cells which could be a way of substituting good genes for deleterious ones in a diseased person. In principle, the way could be opened for curing a sufferer of a genetic disease during his or her lifetime, for the good transplanted cell or gene—and that is what it would be—might be made to multiply normally and supplant the cells containing the bad gene. Such research would seem to be very necessary, but again a great many words have been written about genetic engineering and the spectre has been raised of men engineering men, with science fiction being brought nearer reality. Genetic engineering is in fact very much a basic science at present and there is no possibility of such extreme developments occurring in the foreseeable future. The extent of genetic engineering work is discussed in Chapter 6, so that the present realities of the research can be properly assessed.

Economic aspects

Quite often scientific research discovers a method for curing or providing relief from some disease which in the past has been fatal or has been particularly debilitating. When this is something like the discovery of streptomycin as an effective treatment for tuberculosis, or the development of the Salk and Sabin vaccines for the prevention of poliomyelitis, then few impediments are placed in the way of applying these findings universally to eradicate these diseases. (It is interesting to note, however, that there were a few initial problems with the introduction of the polio vaccine, for one of the first batches turned out to be not completely inactivated and several people contracted polio as a result.) The benefits in economic terms of the eradication of these diseases completely outweigh the costs of development of the treatment and its universal application.

It is not often realized how, in straight economic terms, medical advances can be enormously beneficial. In the United States, it

Biomedical advances: a mixed blessing?

has been estimated that the development of the poliomyelitis vaccine prevented more than 150 000 cases of that disease in the years 1955–61. From the experience of the years before the vaccine was developed, 12 500 of these sufferers would have died and a similar number would have been completely disabled; 36 000 would have been disabled to some extent and 58 000 would have been moderately disabled. The costs of treating these people between 1955 and 1961, it is estimated, would have amounted to $327 million. The cost of the vaccination was about $611 000 and the field trials of the drugs cost about $41 million. It is also estimated that if these people had contracted polio then they would have lost, in lifetime earnings, some $6400 million. So, between 1955 and 1961, the development of the polio vaccine resulted in a net gain for the economy of the United States of $6685 million or about $1000 million a year, taking into account costs of treatment and potential loss of earnings. For comparison, the annual budget of the National Institutes of Health in 1972–3 was $1400 million, and in the same year the British Medical Research Council spent £28·5 millions. Because of the increased costs of medical services, especially hospital treatment in the 1970s, the cost saving now is of the order of $2000 million a year and money will continue to be saved in the future.

Similarly a cost–benefit analysis of the effects of the measles vaccine comes out very much on the positive side, as does an analysis of the effects of the treatment of tuberculosis. Now that these diseases, which at one time were very common, have been virtually eliminated, especially in the developed countries, the social and other effects of eliminating other less common diseases are not so dramatic. And here controversy enters into decisions taken at governmental level on whether or not to institute a programme either to eliminate a disease or to reduce its effects.

Such dilemmas face the decision-makers when scientific developments reach the stage when they can be applied for the relief of suffering. How human suffering can be included in the cost–benefit equation is one of the major questions of today to which there is no clear answer. And when finance is not limitless—which is a fact in all societies—there is the problem of deciding which of two or more potentially life-saving or distress-relieving developments should be supported.

Various sectors of the public whose outlook is clouded by knowledge of an experience of particular diseases will inevitably question governmental decisions when they seem to be taking little notice of some particular disease. With an unlimited supply of money no such problems would arise and all efforts would be made to apply all developments which clearly saved lives and alleviated suffering. Such a Utopia does not exist and decisions of where to spend the available money must inevitably be made.

Since an analysis of the expected benefits must be made it has to be asked whether there is any point in testing antenatally for the presence of a disease which is known to be fatal within a short time of birth. The cardinal purpose of testing is for a diagnosis to be made early enough for the child to be aborted. Clearly there is less to be said for testing if the child is expected to survive for only a short time. On the other hand, although there is no definitive evidence, it is likely that the effects on the mother would be least if the child were not allowed to develop to full term, and this might be an argument for doing tests in the uterus even where an early death of the child is expected. This is unknown ground, however, and there is a need for more research.

Legal questions

The legal problems associated with transplantation are many, as mentioned earlier, but the practice of artificial insemination by donor also runs into legal difficulties. The practice of A.I.D. has largely been allowed to develop without either official support or condemnation, and the present laws in Britain are woefully inadequate to cater for the expansion of A.I.D.

In vitro fertilization does not run into such problems with the law because as a therapeutic practice it is restricted to the sperm of the husband and the ova of his wife. There is no reason to doubt that at some time in the future an infertile husband and an infertile wife could have children in this way, with either or both sperm and ova being obtained from donors. It would therefore be prudent if there were to be legislation covering A.I.D., that it could also take *in vitro* fertilization into consideration.

It is not suggested, of course, that the law should be changed to accommodate every scientific or medical development as it occurs; such a procedure would be impracticable. The courts, and the common law, stand ready to deal with contentious issues, should they arise. But it is generally better to see that they do not arise when difficulty can be forestalled by amendment of a particular law which is both obsolete and a social hindrance. One necessary preliminary to this is for there to be full public debate, fuelled by the appropriate facts from both the scientific and legal sides, whenever science produces developments which could affect society.

A large part of this book deals with issues of infertility. A.I.D., A.I.H., and *in vitro* fertilization are dealt with in detail and the methods available, or soon likely to be available, for alleviating infertility are described. It is therefore most necessary to consider whether or not efforts to alleviate infertility should be supported and the necessary research carried on in the face of what is almost universally felt to be overcrowding and overpopulation of most countries.

The first point to be made is that research into aspects of fertility not only produces information on how to provide children for infertile people but also illuminates other basic scientific facts. For example, research into *in vitro* fertilization, as well as providing children for previously infertile couples, will provide information on the hormone balance in the woman who is a few days into pregnancy. It will also give information on birth defects in a way which might not be possible if the technique of fertilizing human ova in a test-tube had not been developed.

Perhaps the overriding reason for continuing to treat infertility is the basic right of every married couple to have a child, as laid down in the United Nations' universal declaration of human rights. Also, doctors who treat infertility feel that they must do their best to treat patients who come to them, independently of the problems of world population. A limitation on the numbers of children in a family is clearly a far better way of controlling population than a denial of a child to a barren couple. This is an example of the dichotomy between benefits to the individual and benefits to society, which often seem to be in conflict.

2 Artificial insemination by donor and the treatment of infertility

THE artificial insemination of humans first occupied public attention towards the end of the Second World War but, surprising though it may seem, the practice actually stretches back to the end of the eighteenth century. In animals, artificial insemination has a much longer history; there are reports of Arab tribes inseminating their horses as early as the fourteenth century. But it was only in the 1940s and 1950s that the public became aware of, and concerned about, the implications of artificial insemination of women by the semen of donors. As a result, since 1945 three separate bodies have looked into the practice and consequences of artificial insemination in Britain. The last of these reported in April 1973.

At the end of 1945 a Church commission was set up by the then Archbishop of Canterbury and it reported in 1948. It was against A.I.D. In September 1958, a committee was set up by the Secretaries of State for the Home Office and for Scotland under the chairmanship of Lord Feversham; this reported in July 1960. Broadly speaking this committee favoured a *laissez-faire* attitude to A.I.D. It said that A.I.D. should not be prohibited or regulated by law and that it did not favour altering the criminal law to accommodate A.I.D. A recommendation made by the committee to protect a wife who had undergone A.I.D. with her husband's permission from being sued for divorce was not implemented. A further recommendation of the committee was that a wife could be sued for divorce if she had undergone artificial insemination by donor without the consent of the husband. This has not been acted on either, but the Divorce Reform Act of 1969, by substituting the principle of the breakdown of marriage for the principle of the matrimonial offence as the ground for divorce, has made both recommendations, in the form stated, obsolete.

In 1971 the British Medical Association, concerned about an increasing number of requests for information about A.I.D., appointed a panel to look into the medical aspects of human artificial insemination. Sir John Peel, a former president of the Royal College of Obstetricians and Gynaecologists, was the chairman of the panel of six members. Although Sir John's panel recognized that A.I.D. was still a controversial matter it recommended that it should be made available, on a limited basis, within the National Health Service. The panel also adopted a minority recommendation included in Lord Feversham's report which called for the definition of legitimacy to be expanded to include a child born of A.I.D. to which the husband of the mother had consented. Sir John's panel also thought that for the purposes of registration of birth of such a child the husband should be considered to be the father. These recommendations, however, raise the question of how far legal fiction ought to go. Words become steadily more useless if they are stretched to mean what on their face they clearly do not mean.

As the Peel Report recognized, the public attitude to A.I.D. has changed greatly in the past few years. One of the chief reasons no doubt is that many fewer babies are available for adoption, and consequently parents' thoughts turn to other means of having a family. The Association of British Adoption Agencies placed more than 13 700 children for adoption in England, Wales and Scotland in 1967 but it placed only 8417 children in 1971, a drop of almost 40 per cent. This decrease has been accompanied by an increase in the practice of A.I.D. in Britain. Scientific developments in the past few years will, in all probability, make the practice of A.I.D. even more widespread in the future. But these developments create possibilities which go far beyond simply helping childless couples. In particular, there is now the possibility of human sperm being stored for a long time, even many years, at the temperature of liquid nitrogen, $-196\ °C$. Usually, artificial insemination by the sperm of a donor is carried out only when it has been shown that the husband is infertile. But in some instances it has been used when it would have been inadvisable for the husband to father a child; for example, when there is a chance that some rare hereditary disease could be transmitted to the children or through the children to later generations.

As techniques of genetic screening improve, the use of A.I.D. is likely to increase if it becomes possible to pinpoint other diseases which a husband and wife may prudently avoid transmitting to their offspring.

In spite of the possibilities of using artificial insemination in these circumstances, it has in the past, and probably will in the future, be used chiefly to counteract the husband's infertility.

The nature of infertility

Infertility, of course, differs from impotence. An infertile man can deliver semen and accomplish intercourse successfully but his semen is unable to fertilize the egg of the woman. An impotent man on the other hand, is unable to deliver semen although his semen may be quite capable of impregnating a woman. It is the infertile man who might obtain help from A.I.D.

Semen consists of the spermatozoa, which are the seeds for fertilization, within a seminal fluid. Normally, a cubic centimetre of semen contains over a hundred million sperm, but in infertile men the number is often far less and some men produce no sperm at all. These conditions are known as oligospermia, when few sperm are produced, and azoospermia, when none is produced. Some men produce abnormal sperm which can also lead to infertility, but this is rare.

What sperm count is needed before a man can be classified as sterile? One cannot say dogmatically that a man with a sperm count of less than 10 million is sterile, although such a man is extremely unlikely to father a child. But there are documented cases of men eventually becoming fathers who have had very low sperm counts, or who produce no sperm at all when tested. To determine whether a man is infertile or subfertile, several sperm counts are carried out over a space of a few months. If the husband produces one high sperm count during investigation of his infertility, then the doctor is unlikely to proceed with A.I.D.

A source of concern for doctors who practise A.I.D. is in those cases in which a husband fathers a child after his wife first has a child with donor sperm. It is this that makes some doctors treat only women with husbands who produce no sperm at all.

Artificial insemination by husband

Artificial insemination by husband (A.I.H.) has, until now, been used infrequently in Britain, and its chief use has been when husband and wife are both fertile but there are factors which prevent the sperm entering far enough into the cervix. In particular, A.I.H. is used when the husband cannot accomplish intercourse but is able to produce semen by masturbation. A strong case can now be made for increasing the use of A.I.H. when the husband is subfertile because sperm can be frozen and stored for a long time. The probability of a sperm fertilizing an egg depends on both the number produced and on the time between the release of the egg from the ovaries and the act of insemination. To maximize the probability of the egg being fertilized, the sperm of a subfertile man can be collected, concentrated, and stored before being deposited in, or near, the cervix by a syringe at, or shortly before, the time when it is thought that the wife is ovulating. The advantages of this over having regular intercourse during the same few days in the woman's cycle is that more sperm will be delivered at the correct time, thus increasing the chances of fertilization.

The practice of A.I.D. in Britain

What need is there for A.I.D. in Britain? In any discussion on how the technique is likely to develop in the next few years, it is important to know how many infertile couples can be helped.

One of the chief practitioners of artificial insemination in Britain treated 500 couples for infertility by A.I.D. between 1941 and the end of 1971. More than 40 per cent of the women treated became pregnant, several of them more than once, and 298 pregnancies resulted giving 275 live births and 4 stillbirths. Of the live births, 2 children died soon after birth and 3 of the children were abnormal—a casualty rate no higher than that of natural conception.

Reliable statistics on the practice of A.I.D. are not available, only isolated information such as that given above. But according to the report prepared by Lord Feversham and his committee in 1960, one-tenth of all marriages are infertile and probably the husband's infertility is responsible for at least a third of these. In 1972 there were 477 000 marriages in Britain. This could mean that there are 16 000 marriages each year that are infertile because of the husband. The Feversham Report also states that only about one tenth of these husbands are demonstrably infertile, that is, they produce no sperm at all.

Although Lord Feversham and his committee considered that in 33 per cent of infertile marriages the husband is responsible, the range of expert opinion is that this could be anything between 10 per cent and 50 per cent. But only a few couples come and request treatment by A.I.D. Sir John Peel's panel thought that of those who marry each year, some 10 per cent, or 1600 couples, could consider A.I.D. at some time during their marriage.

In 1960 the Feversham Report was able to say boldly that A.I.D. was never undertaken in hospitals and that it was the preserve of the private practitioner. Today, although most artificial inseminations are still carried out in private practice, some are carried out in hospitals under the aegis of the National Health Service. There has been no formal statement during the past ten years allowing N.H.S. doctors to follow this procedure; but if a procedure falls within the limits of normal medical therapeutic practice there is no reason, legal or ethical, for any formal statement to be made about it.

The Peel panel circulated 513 obstetric and gynaecological departments in Britain to ask about the demand for A.I.D., and received 315 replies. A hundred and forty-five departments reported that they had had requests for information about A.I.D. in the previous six months but A.I.D. was undertaken in only ten departments. More than 100 departments, however, referred suitable cases to colleagues —presumably in private practice.

These departments were also asked whether any difficulty could be envisaged if A.I.D. were generally available under the National Health Service in Britain and whether they would be prepared to act as a centre for investigating potential candidates and to undertake A.I.D. Two hundred and thirty-one of the departments did envisage difficulties, but 115 said that they would be prepared to act as centres and 91 departments were prepared to carry out A.I.D. with suitable patients.

Sperm banks

Freezing of sperm will help some infertile couples to have their own children, and as such it is a useful element contributing to medical practice. One consequence of this is that commercial sperm banks might be set up to provide sperm for practitioners of A.I.D. If such an organization became established, it could change the way in

Artificial insemination by donor and treatment of infertility

which A.I.D. is practised in Britain today. The practitioner has until now usually obtained sperm from people he or she knows, whereas if sperm banks become established, then the personal acquaintance which the practitioners hold to be important will no longer be possible.

In the United States, where frozen sperm banks have been in existence for some time, their chief use has been to provide 'insurance' for men who wish to undergo vasectomy. The man, before his operation, deposits sperm in the bank so that in the event of the death of his wife or children he can still have heirs. This raises the bizarre possibility of a man fathering a child after his death. Whether such banks in Britain will be used for the same purposes remains to be determined.

As the current laws and regulations stand in Britain, there is nothing to prevent a sperm bank being set up by private individuals, but its success would no doubt depend on the extent to which husbands about to undergo vasectomy and the practitioners of artificial insemination would make use of it. It is not too early to discuss and decide, as a matter of social policy, whether such banks should have to be registered before they could start business. That such a possibility is not in the distant future can be seen from a report in *Science* (volume 176, p. 632), the weekly journal of the American Association for the Advancement of Science. This report describes the setting up of a personal bank and not a bank to produce sperm for A.I.D.

How do you make a deposit at a sperm bank? At Idant Corporation, a rapidly growing young company that opened a branch in suburban Baltimore, Maryland, a few weeks ago, the procedure is simple.

The customer need only have observed at least 48 hours of prior continence—to ensure a high sperm count—to qualify as a depositor. He strolls into Idant's small laboratory, which is manned only by a secretary and a laboratory biologist, fills out a form, and plunks down the $80 fee required for the processing and freezing of three semen specimens. He then retreats to a tiny room furnished with a comfortable armchair, two pornographic magazines, and an ashtray. (He may also drop off his sample on the way to work, providing it is less than 2 hours old at the time of deposit.) The ejaculate is examined, diluted with a glycerol preservative, and stored in 12 or 15 little plastic vials resembling ball-point pen refills. The vials are stored in three metal canisters and submerged in stainless steel barrels filled with liquid nitrogen, which bubbles away at its boiling point of $-196\ °C$.

The technique of storing sperm at low temperatures is indeed simple. A little glycerol is added to the semen as described above and it is suspended in liquid nitrogen vapour for a short time before being immersed in a tank of liquid nitrogen. All that is then needed is to maintain the level of liquid nitrogen until the semen is used. The technique is in fact so simple that any group practice could store frozen semen. Laboratories would certainly have no difficulty in freezing and storing semen for a long time.

Sperm donors

Sir John Peel and his panel recommended that donors of semen might be paid for the time which they have spent being examined medically, but not, it may be noted, for the semen sample, although in practice there may be little difference. The extent to which semen donors are paid in Britain is unknown and, not surprisingly, information is not easy to come by. Before this practice becomes prevalent in Britain, there is a need to examine the consequences and to comment on its advisability.

A question which has been asked in the past and now, with the issue of payment being raised, will be asked with renewed vigour, is whether donors should as a matter of course be subjected to a series of tests to ensure their fitness, both genetically and otherwise, to donate sperm. The Peel Report states that 'the selection of donors of semen requires care and attention and the panel emphasises that the most thorough examination of prospective donors must be undertaken before they are asked to give specimens'. What sort of examination is needed? Would it be an examination of the donor's mental and physical health, or would it include also an examination of the donor's family history for evidence of genetic disease, or should it require both?

The current practice in Britain is for the practitioner to ask questions of the donor, and the view is taken that, as in most cases the donor is receiving no financial reward, he would have no reason or persuasion to hide the existence of any hereditary defect. If donors are to be paid for their services, however, will the dangers be greater? Sir John Peel mentions in his report that when, in the future, an accredited practitioner of A.I.D. applies to a centre for a sample of frozen semen for insemination he should provide details of the prospective mother and her husband 'so that matching of racial type and blood groups with the donor (can) be undertaken'.

Although the classification of people into major racial groups such as Caucasoids, Negroids, and Mongoloids is clear-cut, the further subdivision of racial types is difficult to define precisely. Race can only exceptionally be established by blood groups, which in any case vary more within racial groups than between them, so that it is hard to see what matching blood groups with the donor would achieve. Behaviour, intelligence, and other characteristics which could be desired by selection are determined by complex interactions between many genetic and environmental factors, and racial differences in these features are, in any case, much more likely to be environmental than genetic. Thus, just what a questionnaire or *curriculum vitae* can tell about the genetic potential of the sperm donor is hard to see.

Even the inheritance of skin colour is not yet clearly mapped out. The only realistic possibility is the screening of the donor for the presence of one or more genes or chromosomal abnormalities which determine genetic diseases and can, at present, be identified. These are, however, a minute proportion of the total possible. Most people carry, in a single dose, one or more genes which, in a double dose (that is, when they are combined with a similar gene on fertilization), could give rise to severe genetic disease. So, for the most part, screening for such genes is simply obtaining information about those people who carry known deleterious genes and it does not mean that others who carry genes which can not yet be identified would be any more suitable as sperm donors. There are so many genetic differences between individuals, and the range of genetic variation that can be produced by any one person is so large, that the value of genetic screening of donors is severely limited—if, indeed, it has any value at all. Certainly the aims of a screening programme must be clearly defined and, in turn, carefully evaluated in terms of present technical possibilities.

With an increased incidence of A.I.D. there will be a greater possibility of marriages occurring between half brothers and sisters who, unknowingly, have a father or sperm donor in common. At the present time such a possibility can confidently be said to be remote because of the relatively few children who have been born as a result of A.I.D. The donors are usually concentrated in a small geographical area but the mothers-to-be come from far and wide. In fact, it is commonly accepted that there is a much greater possibility of marriage between half-siblings (children with one parent in common) who have resulted from extra-marital relationships than between A.I.D. children with the same father. At least there is more chance that the ordinary half-siblings will be concentrated in the same locality.

What are the dangers of brothers and sisters marrying? It is of course illegal in most countries, but it would not be an indictable offence unless the couple knew that they were brother and sister. The limited data which are available on incest show, however, that the genetic dangers although they should not be minimized, may not be as great as is widely feared.

The practitioners of A.I.D. are the first to stress the advantages of artificial insemination over adoption. The child is carried by the mother for nine months and a normal relationship develops between the mother and the foetus. It is claimed that psychologically this is better than adoption for both the parents and the child. In the 1970s, with such concern being expressed about overpopulation, one might argue that the fewer children brought into the world the better and that an infertile couple should therefore adopt rather than undergo A.I.D. Doctors argue, however, that if a woman comes to them needing help and advice on how to become pregnant, it is their duty to help her in every proper way they can; and if this means A.I.D., then overpopulation and other factors should not override. It is also worth noting that the Universal Declaration on Human Rights states that every couple has a right to 'found a family'. The decision should thus be for the woman and her husband to make, but in Britain in the 1970s A.I.D. is not considered an alternative to adoption by most couples. It is possible, however, that this attitude will change. In the United States infertile couples are much more likely to think of A.I.D.

Legal and ethical aspects of A.I.D.

Since 1960 the practice of A.I.D. has gradually increased in Britain, although it is doubtful whether it has increased to the same extent as it has done in the United States in the same period. It is still, however, in spite of the Peel panel's report, considered by sections of the medical profession to be a clandestine operation.

The Feversham committee believed that A.I.D. 'falls within the category of actions known to students of jurisprudence as liberties which, while not prohibited by law, will nevertheless receive no support or encouragement from the law'. It has long been held in the common law that there is a type of conduct which, although regarded as immoral, ought not to be classified as criminal and therefore be punishable by the state. The most common examples are telling a lie (not being a libel or slander), drinking oneself into a state of intoxication in private, fornication, and adultery. The Feversham Committee included A.I.D. in this category.

Has society's view changed to such an extent that the legal position should be re-examined? The Roman Catholic Church has pronounced against A.I.D. in statements from the Holy Office in 1897 and from Pope Pius XII in 1949 and 1956. Its point of view has not changed since the Feversham inquiry. So far as can be ascertained, no Christian church in Britain has explicitly favoured the practice and Jewish Orthodoxy is hostile to the concept. The two bodies set up by the Archbishop of Canterbury, the Commission in 1948 and the Committee in 1960, expressed adverse judgements on A.I.D., although each body had one dissenting member.

Western society, which has been greatly influenced by the Judaeo-Christian religion as well as by the Graeco-Roman tradition of philosophy and law, has emphasized the bond between the begetting and conception of children and the shared or common life of the marriage and the family. Deviations from this nexus were tolerated but they were bound up with other social institutions which have now become discredited, such as the existence of classes of slaves or serfs or social 'inferiors', implicitly excluded from the norms of human relationships.

Within this tradition the conception of a child by sperm from a man who was not the wife's husband was possible only by means of sexual contact between the wife and the father. And such a relationship was denounced as being adulterous.

Fertilization without adultery was not considered in the past because it was not possible. So the fact that artificial insemination with a donor's semen is now possible without adultery invites new ethical judgements.

First, those who hold that the bond between begetting and marriage is inescapable and exclusive will repudiate A.I.D. even though the insemination does not involve an act of adultery. (A judgement passed in 1955, *Dennis* v. *Dennis*, ruled that adultery does not take place unless there is penetration.) But the people who argue along these lines will hold that the semen of a donor will adulterate the marriage even without sexual contact.

Secondly, there are some people whose concept of a bond ends with sexual intercourse between husband and wife. These then regard the semen of another man as a mere fertilizing agent which introduces nothing alien into the marriage, and they would feel free to accept A.I.D. as a remedy for the infertility of their marriage.

What are the motives of the donor? There is a wide realm of uncertainty about these, as there is about the effects of his action upon himself both in the long and short term. The self-gratification in the act of masturbation was for a long time suspect to moralists in the Judaeo-Christian tradition and they would still say that it must be 'ordinate', that the act may be performed only for a proper purpose. In particular the giving of semen for examination, or for A.I.H., is now held to be morally allowable by the Church of England. But giving of semen for A.I.D. is a different matter and moralists of the Judaeo-Christian tradition find it difficult to accord the same freedom to the donor of sperm for A.I.D. that they would to the husband donating sperm for A.I.H.

The moral difficulty here is not so much the pleasure that the donor derives from the act as the requirement that a man should take full responsibility for his own offspring. Christian moralists are not satisfied by the fact that 'someone'—the mother and her husband—will be responsible for the child. It is argued that the potential responsibility is exclusively the donor's.

To become a father, as a donor does, without the continuing father–child relationship, the moralists argue, is to withdraw biological potential from personal potential. This, it is argued, will reverse the long process of evolution by which biological capacities have been 'humanized'. There are cultures where an uncle and not the father assumes responsibility with the mother for bringing up the child, and within these cultures there is a reciprocal relationship in which the father is also an uncle himself and exercises paternal responsibility to nephews of his own. By contrast, the A.I.D. donor is considered a social isolate for he has no continuing responsibility for his offspring.

Some moralists take the word 'love' or 'charity' as the master key to ethical problems and these might argue in favour of A.I.D. by saying that the donor is merely taking the opportunity to do a 'loving act'—by providing a childless marriage with a child. This, they would argue, overrides all other considerations. The case is not a simple open and shut one, however, in that it may be asked what is the nature of 'love' and, in particular, whether it can oblige a man to divorce paternity in the sense of a continuing relationship from his act of donation. He would then, according to one analysis, be performing a less than human act.

There are other problems associated with A.I.D. which are not so problematic. It would seem unethical, on empirical as well as *a priori* grounds, to give a child by A.I.D. either to an unmarried woman or to a wife without the informed and express consent of her husband.

In the same category is the question of whether a fertile husband should ever give semen for the insemination of a woman other than his wife. One essential ingredient of marriage is the mutual and exclusive exchange of procreative powers: a husband may not give to a third party what he has exclusively contracted to give to his wife. It could be argued that the wife might waive her exclusive right by consenting to her husband donating his sperm. But she can do this only if it is already established that the act is not *per se* wrong. If it were a wrong, then consent by the wife would not set it right.

Have the moral and ethical attitudes of society changed in the past decade? The subject is discussed more freely, and with less repugnance, than it was a decade ago; but in spite of the Peel Report there is still no apparent widespread call in Britain for A.I.D. to be provided under the National Health Service.

Legitimacy and A.I.D.

There seems to be no doubt that in law a child conceived by A.I.D. is illegitimate, simply because a legitimate child is defined as the offspring of a married couple. Common law, however, has fully accepted the uncertainties of fatherhood and for this reason it has established the presumption that a child born in lawful wedlock (that is during the subsistence of a marriage or within nine months of the end of the marriage) is legitimate.

The onus of proving that the child is not legitimate rests on the person alleging that it is not and, according to the Family Reform Act of 1969, the alleging person has to show that it is more probable than not that the child is illegitimate.

The law, however, increasingly emphasizes the need for a child to have a settled home with a mother and father figure. This seems to be the outcome of custody and adoption cases in recent years, although of course the rights of the parents weigh heavily with the law.

It seems unlikely in Britain that A.I.D. could be raised in the courts when the husband's permission has been obtained, except possibly in two circumstances: first, in respect to a false entry being made in the birth register; secondly, when claims to titles of honour are at stake.

If the husband, when registering the birth of an A.I.D. child to his wife, enters his own name as the child's father in the register of births, he is committing an offence according to the 1911 Perjury Act. But if the husband believed his statement to be true, he would not be held guilty of the offence. If the space for the father's name is left blank—as it would be if the biological father's (that is the donor's) name is not entered—this indicates that the child is illegitimate. How can this quandary be resolved?

There are a few options open. The register could be marked A.I.D. —as, at present, an entry showing adoption is made. Or a law could be enacted to make inadmissible as evidence in a court of law allegations that A.I.D. had been carried out. In these circumstances, if the husband entered his name as father the register of birth would be incorrect, but the fact that a child was born as a result of A.I.D. could be kept secret.

At present, with an adopted child, the birth register is marked 'adopted' and a separate private register is kept which is cross-referred to the public register. If a similar system were introduced for an A.I.D. child, then the child could always use the short form of the birth certificate if he or she wished to hide the circumstances of his or her birth. If such a procedure were allowed, then the registry of birth would be inaccurate in that the name of the biological father would not be entered. But is this really an important issue? It is clear that for many years husbands have covered up the fact that their wives have borne illegitimate children by procuring birth certificates in their own names. The number of such false entries made on behalf of A.I.D. children will be much smaller than for illegitimate children. Other possible solutions which would not require the setting-up of a separate register would be for the column at present marked 'father' in the birth register to be marked 'father or accepting husband'.

In the past it was difficult, if not impossible, to show unambiguously that the husband was not the father but now that genetic tests are proving more and more reliable, this is not, in general, the case. In the early 1960s it was argued that if the husband's sperm was mixed with that of the donor, or if normal sexual intercourse occurred during the month of the insemination, the husband would not know that he was not the father. The same simple argument cannot be used today. In this respect the law is seriously out of date. As mentioned previously, a child born during lawful wedlock is presumed legitimate unless it is shown that it is more probable than not that the child is illegitimate. A genetic test or series of tests can now show whether the mother and her husband are in fact likely to be the biological parents.

The question to be asked in these circumstances is whether it is advisable to make such genetic tests widely available. Would this lead to much unhappiness and destroy marriages, marriages which would have survived otherwise? All the same, if the question of the parentage of a child came up in the courts of law, is it proper that decisions should be taken based on grounds which are non-scientific and subjective when a simple test can probably settle the question immediately?

In discussing these problems the questions of rights of individuals comes up. If a husband challenges his wife in the courts over the parentage of her child, no one can force the wife and her child to undergo a series of tests. A refusal to undergo such a test if requested by the court, may, however, lead to an adverse comment or conclusion in a paternity suit. (The court will not in fact ask for a blood test if it is thought not to be in the best interests of the child.) The questions to be asked and difficulties to be solved are numerous, and these questions have to be faced and discussed openly. If an analogy has to be drawn in the legal sense then it might be argued that the law relating to the breathalyser and subsequent blood or urine tests has some points of similarity, in that a refusal to provide a sample for analysis after an initial positive test is viewed unfavourably in the courts.

A.I.D. has been raised in divorce cases and it has been ruled, as previously mentioned, that it does not constitute adultery. It is conceivable, however, that it might be raised in different circumstances if the husband's permission had not been granted for the act. According to the 1969 Divorce Act, in Britain this could be a fact relied on in a petition for divorce on the ground that the wife had behaved in such a way that the husband could not be expected to live with her. (This was the concept of cruelty before the 1969 Act was introduced.)

It is conceivable though that A.I.D. could also be raised in a court of law when claims to titles of honour are at stake, but this has not happened as yet. But in wills and settlements, a gift to a 'child' now includes an illegitimate child unless it is explicitly written that this is not so. Similarly, the same wide definition of the word 'child' applies if either parent of an illegitimate child dies intestate. The question to be asked now is whether an A.I.D. child can be considered the illegitimate child of the donor so that these provisions about estates apply to A.I.D. children. Although this has not been tested in the courts, there seems to be little doubt than an A.I.D. child is indeed in law the illegitimate child of the donor, and so could inherit from a father who had played no part in his life other than at its beginning.

Furthermore, on the death of a parent of an illegitimate child that son or daughter is treated as a dependent just as a legitimate child would be. Thus the child can lay claim to part of the estate of the deceased parent—if satisfactory provision has not been made for him. This, presumably, would make the estate of the donor open to claim on the part of that child if he were still a dependent at the time of the donor's death. If an A.I.D. child is accepted as a member of the family by the husband then the husband is responsible for his maintenance, irrespective of whether the child is illegitimate in law.

This extremely unsatisfactory state of affairs has arisen from the fact that the law has in the past ignored the problems of A.I.D. in Britain.

The Feversham Committee recommended in 1960 that a child born as a result of artificial insemination by a donor's sperm should not be considered as a legitimate child of the marriage. But two of the nine committee members disagreed and recommended that the child be accorded the same privileges as a legitimate child. The Peel panel has more recently come out in full agreement with these two dissenting members. No action has been taken, however, and the law is increasingly becoming outdated.

In the United States there have been several conflicting legal decisions on the position of the A.I.D. child. The California Supreme Court in 1969 in a suit for maintenance of a child held that the child conceived by A.I.D. to a married woman with her husband's consent is the legitimate child of the marriage. The State of Oklahoma passed an act in 1967 that made A.I.D. legal and declared that children born as a result of A.I.D. were legitimate. A.I.D. is also legal in Georgia. An intermediate appellate court in New York in 1963, however, ruled an A.I.D. child to be illegitimate.

In 1960 seven members of the Feversham Committee said that:

> If the child of A.I.D. were declared to be legitimate, the consequent devolution of entailed property, titles and dignities of honour might provide an additional motive for resort to A.I.D. This would, of course, be against the interests of those persons who would otherwise have succeeded in default of a legitimate child being born to the husband as a result of normal conception. It would also, we think, be against the interests of society as a whole. Succession through blood descent is an important element of family life and as such is at the basis of our society. On it depend the peerage and other titles of honour, and the Monarchy itself. The Monarchy has not always proceeded by direct descent, but the exceptions only serve to illustrate how much reliance was placed on descent through blood and how much was done to maintain the position that even an apparent usurper should be closely related to the previous Sovereign. The view has been put to us that for a child of A.I.D. and his descendants, to succeed to a hereditary title through his mother's husband would not only be an injustice to persons who were related in blood to the first holder of the title; it would also be an imposition on the Sovereign and a frustration of his original intention in granting a privileged status not merely to an individual, but to a family or group of persons who for ever after could trace a legitimate descent in blood from the grantee. However small the incidence of A.I.D., we do not feel that these considerations ought to be overlooked.

This raises the whole concept of legitimacy. Twice during the past fifty years, in 1926 and 1959, the law has been changed to legitimize children who would otherwise have been illegitimate. The changes, in essence, made it possible for children who were born when their parents were not married to be legitimized provided the parents married each other later. If a marriage was declared void during the gestation period then the children born after dissolution of the marriage would be legitimate. To make an A.I.D. child legitimate throws another issue into the ring. The A.I.D. child is not related by blood to both parents—the vital difference between an A.I.D. child and the children covered by the definition of legitimacy according to the Legitimacy Acts of 1926 and 1959. The majority of the Feversham Committee found this an insuperable barrier to making the A.I.D. child legitimate.

To what extent has society's attitude changed towards legitimacy and illegitimacy since 1960? Parliament in 1969 turned down a clause to the Legitimacy Act which said that 'any child born to a married woman and accepted as one of the family by her husband' should be deemed to be a child of the marriage. Would it do so today?

It can be argued that the concept of legitimacy has outlived its usefulness: the two Legitimacy Acts and subsequent civil and ecclesiastical legislation, and rulings in the courts in the United States, have considerably lessened the social and economic disabilities of the illegitimate child.

A solution would be for legitimacy to be replaced by a new concept of 'acceptance' or 'approbation' which would give the status of social 'filiation' to any child accepted into the family by the husband and wife, no matter how that child was conceived. This would also ensure that all children would have the same rights, privileges, and duties attaching to that status. If there are good reasons for attaching the descent of titles of honour and of hereditary estates to 'blood descent', then they could be treated as special cases and not made the basis of general legislation. A register would then record the social recognition of that child while a separate register could record the child's genetic descent, on the model of the adoption register of today, and would be open to inspection by anyone who could show that he or she had just and compelling interest in its content.

Perhaps the most sensible suggestion made is that the concept of legitimacy be removed altogether. (This would not, however, affect the right of the unmarried woman to apply for maintenance from the father of her child by affiliation, as is done at present.) This would remove all the need for making false entries in the birth register and the provision of an accurate record of genetic descent would give reliable data which would be invaluable in many ways. In particular, if the practice of A.I.D. is to increase and be extended then it would provide material for an objective assessment of the practice which is now, and may be to an even greater extent in the future, a scientific and social necessity. If developments in storing sperm for long periods at low temperatures are commercially exploited in Britain as they have been in the United States, this will undoubtedly increase the pressures for society to take a more definite view of the legal and social implications of A.I.D. Other possible solutions would be for the parents to adopt the A.I.D. child after he was born or for legislation to be passed to put the A.I.D. child on the same footing as an adopted child. The parents could also be given the opportunity of making an application for adoption at the time of A.I.D.

Artificial insemination by donor and treatment of infertility

Summary

The present situation is unsatisfactory in several ways. The child born by the sperm of a donor is illegitimate in the eyes of the law and the husband who hides this fact on the birth certificate is committing perjury. The donor has the same responsibility in law for the child as a father has for an illegitimate child in the usual meaning of the word and, conversely, the A.I.D. child has, in theory, a claim on the estate of the donor.

The conditions in which the practitioners of A.I.D. obtain sperm for insemination, although acceptable in the past, require serious re-evaluation now that the practice is growing. Should sperm banks, which deal in frozen sperm, be allowed to operate without restriction in Britain? And should sperm samples be paid for and the sperm used without seriously inquiring into the genetic and family background of the donor?

The practice of A.I.D. has grown quietly in Britain in the past decade and is now carried out in National Health Hospitals as the Peel Report shows. (The Feversham Committee stated that they were certain that this was not so in the late 1950s.) But it is still a clandestine operation, and in many ways the development of A.I.D. parallels that of sterilization. In 1933 a Committee on Sterilization recommended that sterilization operations should not be carried out. Since then, the legality of a sterilization operation which is necessary for the patient's treatment—that is therapeutic sterilization—seems to be established. The Secretary of the Medical Defence Union in 1966 advised, on behalf of his council, that an operation for sterilization is lawful whether it is carried out on therapeutic or eugenic grounds—that is to prevent any dangerous genetic disease being transmitted to the next generation (see Chapter 4)—or for any other reason, provided that there is a full and valid consent to the operation by the patient. As yet, there has been no such statement made on behalf of A.I.D., but some action is needed to clear the legal inequities surrounding the practice.

3 *Artificial fertilization*

DISCUSSION of artificial insemination waned in the 1960s and it was replaced in the mind of the public by other developments in modern biology which have aroused an even greater reaction. In the past few years A.I.D. has again made news but reports that human eggs have been fertilized and human embryos partly grown in the laboratory outside the body have aroused more interest which has overshadowed discussion of A.I.D. In general, articles on work of this type have not given it wholehearted support. The question which has been raised time and time again is whether such experimental work should be allowed and many unjustified assumptions have been made about its aims. The spectre has also arisen of a baby being fertilized and fully developed outside the womb.

Work of this type is known as *in vitro* fertilization. This term, which is in itself a barrier to communication and understanding between the biologists and the lay public, literally means 'fertilization in glass'. It is used in this context to distinguish between happenings and experiments within the body, which are referred to as *in vivo*, and the experiments and studies *in vitro* which are carried out in the laboratory.

In vitro fertilization has most probably aroused passions because it conjures up a picture of a baby growing in a test-tube. Unfortunately, many of the countless press and magazine articles have done little to divert the public from this line of thought and some articles have even actively encouraged it. But what has not been stressed nearly enough, and in some cases not at all, is that *in vitro* fertilization of human eggs is a complement to artificial insemination by husband (A.I.H.).

Artificial fertilization

Artificial insemination by the husband's sperm is, as discussed in Chapter 2, a therapeutic measure for infertile couples when there is some impediment to the sperm being deposited in or near the cervix of the wife's uterus. *In vitro* fertilization, on the other hand, is potentially a similar therapeutic measure to be used if a wife who has a fertile husband cannot conceive by normal means because some defect prevents the sperm and the ovum from coming into contact.

The primary aim of the biologists working on perfecting the technique of *in vitro* fertilization is to fertilize a wife's ovum outside the body with her husband's sperm and then at the appropriate time in the menstrual cycle to transfer the resulting embryo into her uterus for it to grow and develop normally. The information which this work provides will no doubt have other beneficial effects. A study of the hormone levels which are necessary to achieve successful fertilization and implantation is going to lead to a better understanding of fertility and also offers the possibility of new contraceptive techniques being developed.

For reasons that are mostly unknown, some embryos die in the female reproductive tract at an early stage in their development and are therefore aborted. A natural consequence of studying fertilization outside the body might be that such spontaneous abortion could be better understood. This could have important implications in diagnosing whether or not babies are going to be malformed at birth.

Early work with animals

The transfer of fertilized eggs from one animal to another, in contrast to *in vitro* fertilization, is not a new technique. As long ago as 1890, Walter Heape of the University of Cambridge succeeded in transferring a fertilized egg from an Angora doe rabbit to a Belgian hare doe from which offspring were obtained. Heape described this as follows.

> On April 27, 1890, two ova were obtained from an Angora doe rabbit which had been fertilised by an Angora doe buck 32 hours previously. These ova were immediately transferred to the upper end of the fallopian tubes of a Belgian doe hare which had been fertilised three hours previously by a buck of the same breed as herself. In due course this Belgian hare doe gave birth to six young—four of these resembled herself and her mate while two of them were undoubtedly Angoras.

It was well into the twentieth century before Heape's experiments could be successfully repeated on a regular basis. Since then, fertilized eggs have been successfully transferred in goats, pigs, and cows as well as mice, rats, rabbits, and hamsters. The implications for the animal breeder of this development are enormous for, as with artificial insemination, it is very helpful in selective breeding programmes that are carried out in order to improve the quality and quantity of livestock.

In 1954, fertilized sheep eggs were sent from the United States to Cambridge, England in a vacuum flask, where they were transferred to the uterus of a foster sheep and subsequently developed normally. Yet another way of transporting sheep's eggs was first used in the 1960s when C. E. Adams, together with a pioneer in this field I. E. A. Rowson of the University of Cambridge, sent fertilized sheep's eggs to South Africa in what was undoubtedly a novel container: the oviduct of a rabbit. Once the rabbit arrived in South Africa, the eggs were transferred to the uterus of a foster sheep and eventually normal lambs characteristic of the donor were born. The cost of transferring the sheep across the world amounted to no more than 50p for each egg, considerably less than the cost of transporting the sheep by conventional means. In future, when deep-frozen embryos, stored in liquid nitrogen in the same way as sperm, can routinely be transferred from place to place the cost will probably be even lower. But the improved convenience of the procedure will be as important as the lower cost.

The first successful transfer of fertilized animal eggs was reported in 1890, but it was not until 1954 that the first successful fertilization of animal eggs *in vitro* was reported by Thibault and his colleagues in France. Their success was achieved with rabbits. Five years later Chang and his colleagues from the U.S.A. reported that they had successfully transferred eggs fertilized *in vitro* to the oviducts, giving rise to the birth of young rabbits. Sheep eggs were first successfully fertilized *in vitro* by Dauzier and Thibault in 1959, but even in the 1970s the success rate of implantation of sheep eggs is not as high as it is for other animals.

Research on humans

What are the facts about infertility that make the successful development of *in vitro* fertilization particularly important as a therapeutic measure in humans?

In Western Europe and the United States it is estimated that 10 per cent of marriages are infertile for one reason or other. Only in the past few years has it been realized that a high percentage of these marriages, possibly as many as a half, are infertile because of lesions which affect the pelvic organs of the wife. About a fifth of this total could possibly be made fertile by surgery which makes the Fallopian tubes accessible to the ova released by the ovaries, but this still leaves a large number of women who have no chance of becoming pregnant by normal means because their Fallopian tubes are hopelessly blocked.

It has been estimated that there are some 20 000 women in Britain alone who are infertile because of such lesions. In the United States the number could be as high as a million. These women have at least one healthy ovary and a healthy uterus but no means of channelling the ova between the two. One of the first attempts to help these women overcome their infertility was made in 1911 by Estes, who partially transplanted an ovary into the uterus in the hope that normal conception would take place there. Unfortunately, such operations have rarely been successful.

With this background, Dr. R. G. Edwards of the Physiological Laboratory at the University of Cambridge started to collaborate with Mr. P. C. Steptoe, a surgeon at the Oldham and District General Hospital. Taking as their cue successful work with mice they set out to help these women by completely circumventing the Fallopian tubes. They planned to remove the ova from the ovaries at the appropriate time in the menstrual cycle and, in the laboratory, faithfully reproduce the events which occur between the time the ovary releases the ovum to the time of fertilization by the sperm. The stage is then reached following fertilization when the embryo, if it were inside a woman, would be within 2 or 3 days of attaching itself to the wall of the uterus (called implantation) that is, some 3 or 4 days after fertilization. Their scheme would then be completed with the transfer of the embryo to the uterus to grow and develop normally.

The first stage of the work consists of investigating the condition and the position of the ovaries in the patient. In this study, called laparoscopy, a gas is introduced into the abdomen to separate the organs so that they can be clearly identified. Steptoe and his colleagues at Oldham have found on the basis of such observations that about 13 per cent of their patients need a preliminary operation in order to make the ovaries accessible so that ova can be removed. Only after this preliminary search can a further laparoscopy be carried out to remove the ova if required. More than 5000 laparoscopies have been carried out at the Oldham General Hospital since 1959 with no deaths attributable to laparoscopy and with minimal side effects. One death was due solely to anaesthesia, a risk estimated at 1 in 10 000 of all general anaesthetics. The woman usually recovers rapidly from the laparoscopy operation, leaving hospital within 24 hours, and is back at her normal tasks within 2 to 3 days.

To ensure that the laparoscopy is successful and that a sufficient number of ova are recovered the woman is treated with a class of sex hormones called gonadotrophins, which cause the ovary to produce several ova simultaneously. The extra ova so produced would normally not be used by the woman, for during her reproductive years she releases only about 500 out of some 500 000 ova formed before birth.

During the normal passage of an ovum through the Fallopian tubes, it undergoes a series of changes called maturation, which have started in the ovary and are caused by the hormones in both the ovaries and in the Fallopian tubes. Thus a medium had to be developed in the laboratory in which the ova would mature just as if they had undergone a normal passage between the ovary and the tubes. Similarly, sperm undergo a kind of maturation within the woman before one of them can penetrate and fertilize an ovum. This process also had to be reproduced in the laboratory.

Animal ova are known to divide in artificial culture even without having been fertilized by sperm. It is thus essential to be able to identify a sperm within the ovum at the start of fertilization, and residual parts of the sperm in the divided ovum afterwards, before it can be said that fertilization has occurred. Edwards and Steptoe reported these observations in 1969 and 1970 and so confirmed that fertilization had indeed taken place.

In 1971 Edwards and colleagues reported that they had succeeded in growing a human embryo in the laboratory until it had divided into 100 cells or so (the so-called blastocyst stage) at which the embryo, in normal circumstances, becomes attached to the wall of the uterus. At this stage the cell mass is made up of an outer shell which eventually becomes the placenta and an inner cell mass which eventually develops to become the foetus (see Pls. 1 and 2).

At the same time as Edwards and Steptoe reported their results, L. R. B. Shettles of the State University of New York reported that he had also grown an embryo to a similar stage of development outside the body.

Edwards and Steptoe have been attempting to implant such week-old embryos in the mother's uterus since the beginning of 1972 but with no success reported up to the end of 1973. This is because several factors need to be clarified before a successful implantation can occur.

First, for the implantation to be successful, the hormone levels within a woman who is a week into pregnancy may have to be artificially induced; but the hormone balance at this point is poorly understood. The timing of the transfer is also critical.

Of particular concern to Edwards and Steptoe and the other scientists who are studying *in vitro* fertilization of human ova is the possibility that the child so born will in some way be defective as a result of the way the early embryo has been handled.

Edwards and Steptoe are quick to point out that when they do succeed in implanting a blastocyst in the uterus the development of the embryo will be carefully monitored in order to observe any possible abnormalities. Such surveillance will include ultrasonic investigation early in pregnancy, followed later by studies of cells removed from the fluid which surrounds the foetus in order to determine whether the foetus is developing normally (see Chapter 4 for details).

Animal experiments and birth defects

How successful have similar experiments been in animals? In recent years many scientific reports have been written describing the successful outcome of experiments involving the transfer of embryos in animals. In some of these experiments, ova have been transferred which have been fertilized *in vitro*, but most of the experiments concern the transfer of fertilized ova obtained directly from the ovaries.

These transfers have been made to the uterus, the natural place for growth to occur, or to the oviducts (Fallopian tubes) when there was a need for the fertilized ova to mature before implantation. Frequently, experiments have been carried out which have interposed additional stages between obtaining a fertilized ovum and implanting it. These intermediate stages include maturation *in vitro*, as described above for the work of Edwards and Steptoe for humans, storage of the egg, manipulation, and even irradiation. The extent of such work in animals can be appreciated from a review article published in 1971 which referred to 451 original papers written by biologists on the transfer of embryos, 15 theses, and 41 books and reviews dealing with the same subject.

This wealth of information, which includes data on the death of animal foetuses and the incidence of abnormalities in these foetuses, enables quite firm conclusions to be drawn about the success of transfer experiments. All these reports show that there is a high probability of achieving a successful embryo transfer in certain animals. With the knowledge gained in recent years on the hormone levels needed in the recipient female at the time of implantation, pregnancy and the birth of young now generally follow such experiments in animals.

Significantly, in these animal experiments, birth defects following the transfer of fertilized ova are rare. Indeed there does not appear to be a single report in the scientific literature in which any of the standard procedures (including ovulation, recovery of the ovum, maturing the ovum *in vitro*, fertilizing the ovum *in vitro*, and culture in some artificial medium followed by transfer) is significantly linked with an increase of birth defects above the normal level, which lies between 1 and 5 per cent of all births. The lack of abnormalities in the young born after such transplant procedures does not imply, of course, that there were no prenatal losses. In fact such losses turn out to be higher than in natural conditions, but the losses occur generally before the stage at which the embryo attaches to the wall of the uterus.

The implications of the absence of any observed increase in birth defects over and above those normally found when fertilized ova are transferred are brought into focus when experiments are carried out which deliberately set out to induce birth defects in the young.

Pre-implantation embryos of animals have been exposed to extreme temperatures, ranging between $0\,°C$ and $-269\,°C$; they have been irradiated with gamma rays from radioactive isotopes of cobalt; a single cell has been separated from embryos of 2 cells, 4 cells, and 8 cells before being allowed to grow by itself and then transferred; sheep and cow embryos have been transferred to the uteri of rabbits for several days before being inserted in sheep or cows; a part of the outer shell of the blastocyst (called the trophoblast layer) has been removed; and the blastocyst has been dissected to form two blastocysts which have then both grown normally when transferred to a uterus. The inner cell masses of blastocysts have also been interchanged without any ill effects on the resulting embryos; blastocysts have been augmented by the injection of cells from other blastocysts; embryos have been stored for 30 minutes at $-79\,°C$. Most recently,

Artificial fertilization

embryos were frozen at a temperature of −196 °C for several months before being implanted. None of these procedures has resulted in any significant increase in birth defects, a truly remarkable testimony to the resilience of the embryo during the first stage of its life. And the reasons for this are well understood, as explained later.

There have, though, been a few instances of scientists claiming success in experiments designed to induce teratogenesis (production of congenital abnormalities) at the pre-implantation stage but the results of these experiments have been far from convincing. The results may have been affected indirectly by the mother animal suffering a change in her environment at a later stage in her pregnancy. Werthermann and Reiniger reported in 1959 that they had observed defects in the eyes of young rats whose mothers had been deprived of oxygen in the first week of pregnancy. Audrey Smith in 1957 froze hamster foetuses for 30 to 55 minutes on different days of pregnancy up to the twelfth day and then killed them just before the end of the pregnancy. She found that in a litter from an animal treated for 45 minutes at $2\frac{1}{2}$ days of gestation all the foetuses were 'grossly deformed', but significantly, no other animals frozen during the pre-implantation period gave such a result. In 1959 Rugh and Grupp described an increase in certain abnormalities in the young born to mice which had been irradiated between $\frac{1}{2}$ and $4\frac{1}{2}$ days after ovulation. Unfortunately, Rugh and Grupp did not have proper control animals—that is, similar mice which had not been irradiated. This investigation was repeated in 1968 and no such effect was found.

It must be noted that the resilience of the embryo to maltreatment applies only at the pre-implantation stage. The response of embryos after they have attached themselves to the wall of the uterus is completely different. Once this occurs, the embryo is said to be in the organogenesis stage, that is the stage when the heart, lungs, liver, limbs, and other organs are being formed. At this stage, a wide range of agents has been shown to cause birth defects, including irradiation, chemical mutagens, thalidomide, hormones, vitamins A and D, rubella, dietary deficiencies, and deprivation of oxygen to the foetus. Well-trusted agents such as penicillin, streptomycin, insulin, cortisone, and even aspirin have also been shown to be teratogenic, that is, causing birth defects at this stage of development. In some instances these agents have even been shown to produce birth defects which have affected the entire litter.

Why, then, is the embryo apparently immune from damage during its pre-implantation stage, that is, up to and including the blastocyst stage, and so heavily affected afterwards? The explanation may be that during the early stage of an embryo's development all the cells appear to possess, in large measure, the capacity to form any part or even the whole of the future individual. Thus any cell which is lost or killed at this stage can be compensated for by the growth of another cell. As the chief effects of a teratogenic agent are to kill cells or prevent their multiplication, the behaviour of the embryo after being subjected to one of these agents is dependent on the extent of the damage done. Above a certain level of damage, the embryo dies, whereas below that level the deficiency is made up by the growth of new cells. This could explain the higher incidence of deaths and the lack of congenital abnormalities in embryos before the implantation stage.

Once implantation occurs, the cells of the embryo begin to divide at a fast pace and at this stage different cells are set aside to form the various tissues and organs. Cell death during this organogenesis cannot be made good, because once this stage is entered all the cells of the embryo have been assigned to various parts of the body and cannot be switched to making any other tissue or organ. The results of cell death or damage are therefore permanent structural and functional defects in the young animals.

Potential genetic hazards

Diseases which result from defects of the chromosomes are discussed in Chapter 4, but there is one particular defect which might arise with *in vitro* fertilization more often than with normal births. This is a defect where the fertilized egg contains an extra full set of chromosomes over and above its normal content.

Normal human cells have 23 chromosome pairs, that is 46 chromosomes in all; but a sex cell, whether sperm or ovum, contains only 23 chromosomes, one half of the full set. A cell formed by the union of a sperm and an ovum will then normally contain a full complement of 46 chromosomes. If, for example, two sperms fuse with an ovum simultaneously, then the fertilized ovum will contain 69 chromosomes instead of the normal 46.

It is estimated that about 1 per cent of natural conceptions have an abnormal number of sets of chromosomes, formed either by two sperms fusing with an ovum simultaneously, as described above, or alternatively because, through some other defect, the sex cells them-

Artificial fertilization

selves prove to have an abnormal number of chromosome sets. In most cases, any foetus that starts to develop with this abnormal number of chromosomes will abort naturally very early in pregnancy. The situation where two sperms simultaneously fertilize an egg might occur more frequently with *in vitro* fertilization. It is more than likely, however, that any such foetus after being placed in the uterus to grow would be aborted spontaneously.

The foetus resulting from *in vitro* fertilization could still suffer from some genetically determined disease resulting from a trait inherited from the father or mother. These defects, which are discussed in Chapter 4, do not result from an abnormal number of chromosomes except in special cases, such as Down's syndrome (mongolism), and, would in fact manifest themselves whether the foetus is conceived naturally, or following *in vitro* fertilization (or any other process of manipulation of the ovum or sperm).

The term 'test-tube baby' has not been mentioned so far. This might seem strange in a chapter which is apparently exclusively concerned with test-tube babies, but the avoidance of the phrase is deliberate. The reason is partly that the scientists working in the field of fertility research do not like the term. They feel, with justification, that it is what this term conjures up in the imagination of the general public that has done most harm in the past few years.

But what *does* the public think when it hears of test-tube babies? No doubt most would think of A.I.D., but many would envisage the 'demon scientist' carefully nurturing a child in a glass womb, completely devoid of any contact with a mother. The truth lies a long way from this. It should be emphasized that no scientist has grown a human embryo in a test-tube, or any other piece of laboratory equipment for that matter, for longer than 7 or 8 days.

It may well be asked at this point whether it is indeed possible to develop a baby completely outside the womb and whether some scientist could, and would, attempt to grow an embryo to full maturity without the woman who contributed the ovum having the satisfaction of becoming a mother in the usual sense of the word. This is not possible now, but many scientists involved in ovum transplantation research would not deny that, given sufficient time, money, and effort, it might be possible at some time in the future.

There is certainly a record of one scientist who would attempt to grow a baby to full maturity outside the human body, if given the chance. The record does not allow this American to be identified but he is quoted in the book *The Second Genesis* by Albert Rosenfeld as follows: 'If I can carry a baby all the way through to birth *in vitro*, I certainly plan to do it—though obviously I am not going to succeed on the first attempt or even the twentieth.'

Ethical questions

Adequate precautions are available to ensure that research towards the goal of a complete test-tube baby will not be started without the problem being thoroughly discussed. But research on related aspects will undoubtedly and rightly be supported. Who will deny that there is a need to develop incubators to ensure that premature babies have the greatest chances of life? Should a line be drawn, however, at say, 6 or at 4 months at helping a foetus who suddenly has to face life outside the uterus, perhaps because of some unfortunate accident to its mother? And if an incubator were developed for a baby of 4 months then should a line be drawn and help refused to even younger foetuses? The earlier the baby has to leave the womb, the greater the chance of its being physically or mentally handicapped.

Assistance for the premature baby would, by most, be considered one of the basic duties of society. Such advances also lead inexorably towards the complete development of a baby outside the body. Surely, research on maintaining the premature baby should not be given low priority merely because it also leads towards the possibility of developing a baby from fertilization without having been inside a woman's womb?

How are decisions reached on whether or not to proceed with a line of research if ethical principles are at stake? In Britain the chief responsibility for medical ethics lies with the experimenter, his profession, if any, his employer, and the local ethical committee which has to approve any line of research. The next stage is reached when an organization such as the Medical Research Council is asked to grant money for the research. In general the Council weighs the scientific interest and probability of success of any application for support which it receives against the costs involved and for the most part these are the only factors taken into consideration. The ethics of all proposals are considered where necessary by the Medical Research Council's research boards.

What specifically are the fears of the layman? These are probably neatly summed up in the summary of an article that appeared in the magazine *Nova* in May 1972. 'The Perfect Baby or the Biggest Threat

Since the Atom Bomb?' ran the caption on the cover. The summary read as follows.

If today we do not accept the responsibility for directing the biologist, tomorrow we may pay a bitter price—the loss of free choice and, with it, our humanity. We don't have much time left. The questions raised by genetic engineering are every bit as crucial to the future of the human race as those raised by the splitting of the atom. The making of the atom bomb was surrounded by secrecy, as it happens by war-time security needs. However, research into reproduction, the genetic code and how to use it, is going on in laboratories around the world. Nobody's national security is at stake. Yet there is secrecy. The results may face mankind with its most agonising quandaries ever. Can we remain essentially human through the next few generations? Are we mature enough to face the Faustian prospect of actually determining the quality of our descendants? Huge questions indeed. But too many scientists engaged on the pioneering experiments maintain silence towards, if not contempt for, the general public. We must learn that to make adult moral decisions people must be given the facts on which to base their own judgements.

But the medical profession is very much aware of the implications of the work on *in vitro* fertilization. For one thing, this book is a testimony to this fact and no secrets are being withheld by the scientific and medical communities as far as it can be ascertained. The British Medical Association through its Board of Science and Education produced in March 1972 a document of professional standards as a basis for discussion within the profession. The panel which produced the report saw no objection to the experimental research carried out so far on the culture of human fertilized ova but considered that the implications of this field of research should be kept under review. The panel also pointed out that there is no legislation governing experimental work on *in vitro* fertilization. Although it is important to keep options open on developments in this field of research, safeguards should be established, and the panel expressed a willingness to help formulate these guidelines for the medical profession.

The panel also said that it is important that patients with problems of infertility who are considered suitable for *in vitro* fertilization studies should be given detailed explanations of the full procedure and implications before any experimentation involving *in vitro* fertilization is undertaken. An undertaking should also be given that only the husband's sperm will be used for fertilization of the ova removed from the wife's ovaries. The panel also stressed that it would be unethical to use a foster uterus, that is, to transfer a fertilized ovum from one woman to another woman's uterus.

As mentioned earlier in this chapter, it may be practicable to diagnose certain abnormalities of the foetus in fertilized ova at an early stage of development, before they are implanted into the uterus for normal growth. The panel said that provided such techniques could be developed without damaging the ova then screening for abnormalities of the foetus in this way would be preferable to terminating the pregnancy at a much later stage.

In this statement produced by the B.M.A. panel several of the anxieties expressed in articles for the lay public are given consideration. It has been mentioned that the technique of *in vitro* fertilization once fully developed could be exploited and not restricted to a husband and wife. What of the woman who wishes to have a child but for health reasons, or because she has an absorbing career, or perhaps even on the grounds of vanity, does not wish to bear it herself? With *in vitro* fertilization as a viable technique this woman's fertilized ovum could be placed in a foster uterus to grow and develop. Should this be allowed to happen in spite of the strictures of the B.M.A.?

There are certain circumstances in which the growing of a foetus in a foster uterus might be considered allowable. As mentioned earlier the chief aim of the scientists working in infertility research is to allow women who have a normal uterus and healthy ovaries to bear children if the oviducts, or Fallopian tubes, which connect the uterus and the ovaries are not able to channel the ova between the two. But what of the woman who has one healthy ovary but no uterus, for example? Should her suitably fertilized egg be allowed to grow in another woman's uterus? Or conversely, what of the woman who has a healthy uterus but does not produce ova? Should she be deprived of the joy of bearing a child especially if it would be possible for her to do so with an ovum taken from another woman which had been fertilized, *in vitro*, with her husband's sperm? This is the exact complement of A.I.D.

The possibilities do not end here. There is the case of the infertile husband and the infertile wife. In this situation both the fertilized ovum to be transferred to the wife's uterus and the sperm could be obtained from donors, and the ovum either fertilized *in vitro* or possibly fertilized naturally before being removed from the donor's Fallopian tubes or uterus.

These are the possibilities which the British Medical Association's Board of Education and Science in 1972 decided should not be encouraged. They will, of course, not become serious possibilities

Artificial fertilization

until *in vitro* fertilization and successful implantation occur, but these issues are bound to be raised as soon as this happens.

In vitro fertilization, implantation, and subsequent birth of a normal child, who is the genetic offspring of his parents poses no overt legal problems, in contrast to those associated with the birth of a normal child after A.I.D. Claims for compensation because of the negligence of the doctor or scientist could arise as a consequence. And these claims could arise in spite of all the care taken in the fertilization and implantation procedures merely because some genetic disease appears in the child which is carried by, but is not evident in, the parents. Liability, of course, will not be automatic and in such a case it must be shown that the child's disease arose out of the manipulation procedures. The burden of proving that the manipulations carried out by the doctors or scientists had caused the damage to the child lies with the plaintiff in Britain, as in most countries. In the United States, however, in certain circumstances the burden of proving that the damage was not caused by the defendant can fall upon the defendant. These circumstances arise in cases of 'ultra-hazardous activities', but what this term embraces is debatable.

Legislation to protect the early embryo has been conspicuous by its absence in Britain, but a working paper introduced by the Law Commission 'Injuries to the unborn child', opened up the issue in Britain in 1973. In due course this should lead to legislation. In the absence of legislation, however, the cases which have arisen have been common law suits by deformed or defective children demanding compensation for negligence. A case which could be relevant to any future case arising out of negligence during *in vitro* fertilization occurred in 1967 when a doctor was sued for negligence by a child and its parents. The mother suffered from German measles early in pregnancy and according to her the doctor advised against abortion. The child was born defective. The courts, however, turned down the claim, ruling that since the only alternative was abortion, it was unable to decide between the merits of non-existence and existence in a deformed state. Justice Weintraub stated the court's dilemma as follows: 'Ultimately the infant's complaint is that he would be better off not to have been born. Man who knows nothing of death or nothingness, cannot possibly know whether that is so.' The weight of legal opinion since then supports this decision.

There is also no statute in England and Wales which specifically controls the conditions in which experimental operations can be carried out on living persons. The law therefore places a heavy burden of proof on the medical profession to show that the procedures have a therapeutic value, if the matter is raised in the courts.

What are the ethics of fertilization *in vitro*? The issue is rather clouded because as far as humans are concerned the technique is still very much in the research stage. So in one sense the research element in the work falls under the general heading of 'non-therapeutic clinical research', as defined in statements on responsibility and research (e.g. the Helsinki Declaration 1964), since there is no patient to consider until the act of fertilization has brought one into existence. But if fertilization *in vitro* becomes a prelude to implantation then, of course, the woman who is possibly to become a mother becomes a 'patient' and both she and the embryo become the objects of professional clinical care.

The crucial question, however, is what of the embryo before implantation. What is its status before implantation? The blastocyst has, of course, the full genetic potential for becoming a human being and will become one if implantation and gestation are successful. At what stage of development should the status of a patient be attributed to the embryo or foetus? That is, when should the embryo receive the same ethical safeguards and protection which are accorded to human beings in research, combined with full professional care? Also, how should the degrees of acceptable or unacceptable risk be assessed? If a blastocyst is regarded as no more than a disposable tissue, then high risks can be taken because the tissue could be jettisoned if harm resulted. But if the full potential for growing into an adult human being is considered to entitle the blastocyst to some degree of the protection ethically accorded to human beings, then limitations are at once placed upon all experiments.

These are the difficulties, and clearly ways of approaching them differ. The Roman Catholic Church, for example, considers that human life begins at fertilization, while others consider it to begin when the embryo is implanted in the womb. Many people would disagree with both these considerations and would argue that human life begins at a much later stage of pregnancy.

4 Genetic screening and selective abortion

ESTIMATES vary of the number of children born in Britain each year suffering from severe genetic diseases. A common figure that is quoted is 2 per cent of all births, which amounts to 16 000 children a year, but a more realistic and verifiable figure is probably 0·4 to 0·5 per cent of births: 3000 to 4000 children a year (see Tables 4.1–4). Many more children, however, have various genetic predispositions. Genetic diseases are also the cause of some 11 per cent of deaths of children in Britain: 16 000 infants died in Britain in 1971, about 1700 of them from genetic diseases.

In the past few years an increasing number of diseases of children have been diagnosed before birth. The stage of pregnancy varies at which the diseases can be said to be present, but several diseases can now be detected early enough for an abortion to be carried out. In Britain, the 1967 Abortion Act allows mothers who carry a substantial risk of bearing a seriously handicapped child the opportunity of having the pregnancy terminated. With some diseases, however, a diagnosis made before birth, even though it may be made too late to abort the foetus, might in principle be of help in deciding what treatment to introduce after birth.

In spite of the advances which have been made in the understanding and diagnosis of these diseases, most highly disabling genetic or congenital diseases cannot yet be detected before birth. For the diseases which cannot be detected during pregnancy, and for which early treatment is essential to nullify their detrimental effects, a diagnosis soon after birth is desirable. Such remedial action can lead to a longer life-expectancy and a lessening of the suffering for the child and its family.

Population-wide schemes for the detection of some diseases after birth exist in parts of Britain, Europe, and North America. For example, over 80 per cent of children born in Britain are tested within a few days of birth by means of a simple blood test to see whether they are suffering from the disease phenylketonuria, which, as discussed later in this chapter, usually leads to severe mental retardation if not treated.

The defects which can now be detected while the foetus is still at an early stage of development are mostly caused by defective genes or by abnormalities in the chromosomes or in the number of chromosomes in the cells of the foetus. This is not to say that other diseases have not manifested themselves at such an early stage of development, but rather that procedures for studying diseases prenatally are so far limited to analysis of the amniotic fluid which surrounds the foetus within the uterus. A sample of the amniotic fluid is removed from around the foetus by amniocentesis and diagnosis are made from analyses of this fluid. The success of the procedure is based on the fact that the foetus sheds cells into this fluid, and these cells can be studied to establish whether or not the foetus carries a particular genetic trait.

There are three main classes of genetic diseases: those which are due to aberrations in the number of chromosomes or gross changes in particular chromosomes; those defects due to particular genes which make up the chromosomes; and 'polygenic' diseases, where more than one and perhaps many genes are affected.

Chromosomal diseases

Human cells normally have 23 pairs of chromosomes, each chromosome being made up of at least several tens of thousands of genes (Pl. 3(a)). Some diseases arise when the number of chromosomes in the cells is different from 46, although in a few rare cases normal people can have only 45 chromosomes.

Most cases of mongolism or Down's syndrome, for example, arise when there are three number 21 chromosomes in each cell instead of the normal two. In about 3 to 5 per cent of mongols, however, the number of chromosomes is normal and the disease manifests itself because a part of one chromosome has broken off and attached itself to another. Most common chromosomal diseases are, in fact, caused by a variation in the number of chromosomes rather than by aberrations of the chromosomes themselves.

Out of a total of 800 000 or so births every year, about 880 mongols are born in Britain (an incidence of about one birth in every 900). In the United States, more than 7000 mongols are born every year.

Single genes and the mechanics of inheritance

The way in which diseases caused by defects of the genes arise can be understood by first examining the way in which cells grow, and how genes are transmitted from parent to offspring.

TABLE 4.1

Annual frequency of chromosomal diseases in Britain (assuming 800 000 births a year)

	Approximate number of births per year (percentages of births in parentheses)	
Autosomes		
Down's syndrome or mongolism (trisomy 21)	880	(0·11)
Other trisomies	160	(0·02)
Abnormalities and variants	1280	(0·16)
Other mixed abnormalities	160	(0·02)
Total	2480	(0·31)
Sex chromosomes		
XO, Turner's syndrome (Only one sex chromosome, abnormally developed sterile female)	160	(0·02)
XXY, Klinefelter's syndrome (sterile male, though often not very abnormal)	560	(0·07)
XXX, trisomy for X (normal female, possibly with reduced fertility)	320	(0·04)
XYY (males, mostly normal)	720	(0·09)
Other mixed abnormalities	64	(0·008)
Total	1824	(0·23)
Overall total	4304	(0·54)

Only those abnormalities found at birth are shown in the table. Other types of abnormalities with severer effects are found in early spontaneous abortions and are a major cause of such early abortions. Some of the abnormalities found at birth, such as XO, Turner's syndrome, occur with much higher frequencies in spontaneous abortions, showing that they may have a severe effect on the early development of the foetus. Most of the individuals with autosomal chromosomal diseases other than Down's syndrome die soon after birth. The same is not true for the sex chromosome diseases, though here fortunately only the least frequent disease (XO, Turner's) is associated with consistent severe detrimental effects. Data based in part on a publication by Dr. P. Jacobs

48 *Genetic screening and selective abortion*

Cells can divide in one of two ways. In normal cell division, known as mitosis, two identical cells are formed from the original cell and the two cells so formed each contain 46 chromosomes. Sex cells, on the other hand (sperm and ova), are formed from ordinary cells by a process of cell division known as meiosis, in which the final cells contain only half the number of chromosomes of the parent cell. That is, each sex cell contains 23 chromosomes, one half of the full set. Thus when two sex cells unite, the cell so formed once again contains the normal complement of 46 chromosomes, half of these obtained from the mother and half from the father.

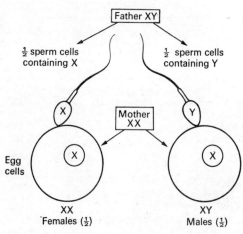

FIG. 4.1. Normal fertilization: formation of equal numbers of males and females. Only the sex chromosome is indicated in each sperm and in the nucleus of each egg cell.

The sex of an offspring is determined by what are known as the X and Y chromosomes. Each male cell, except the sperm cells, contains an X and Y chromosome, whereas female cells, ova apart, have two X chromosomes. The sex of any child is determined by the combination of X and Y chromosomes that occurs after an ovum and sperm have united (Fig. 4.1). There are two types of sperm formed by meiosis, one containing an X chromosome and the other containing a Y chromosome. All ova, on the other hand contain one X chromosome. The two types of sperm are produced in equal numbers by the man and so there is an equal chance that a sperm containing an X chromosome or a Y chromosome will ferti-

Genetic screening and selective abortion

lize any ovum, resulting respectively in a female (XX) or a male (XY) embryo. In practice, it is thought that conditions may not always quite favour the X and Y sperm equally and this, in part, may account for the fact that the number of males born is generally a few per cent more than the number of females.

Some diseases, for example, Duchenne muscular dystrophy, the Hunter syndrome, the Lesch–Nyhan syndrome, and haemophilia, are caused by abnormal genes on the X chromosome (see Table 4.2).

TABLE 4.2
Expected frequency of occurrence in Britain of diseases linked with the sex chromosome (assuming 800 000 births a year)

	Approximate number of births a year
Duchenne's muscular dystrophy	160
Haemophilia (both forms)	80
All others (including some severe mental defects)	160
Total	400

The Hunter and Lesch–Nyhan syndromes, which are sex-linked diseases that can be detected in the uterus, occur with very low frequencies, of the order of a few births in every million.

Duchenne muscular dystrophy is a disease which comes on when the child is between 2 and 6 years old. The first signs are a weakening of the muscles around the pelvis. Weakness then gradually extends to all other parts of the body and the child is unable to get up without turning to one side. Gradually the child weakens and few children survive their teens.

The Hunter syndrome is marked by a progressive physical and mental retardation leading to death, usually from heart failure, during the third or fourth decade of life.

The Lesch–Nyhan syndrome is characterized by mental retardation, cerebral palsy, and a compulsive biting of the lips and finger tips.

Haemophilia is a disease where there is an inborn tendency to bleed, owing to a defect in the clotting power of the blood. It has been known for a long time, and occurred in a number of the descendants of Queen Victoria. In fact, its pattern of inheritance, which it shares with the other diseases connected with genes on the X chromosome, was recognized at least 1300 years ago. According to the Talmud, boys whose brothers were bleeders were exempt from circumcision, as also were the sons of sisters of women who had had male bleeder children. The exemption, however, did not extent to fathers' sons by other women, making it clear that the inheritance pattern of the disease was already understood at that time.

Only men can in general have these diseases though they are transmitted through the female side. This can be seen as follows. A normal male has the genetic make-up XY, while the normal female is XX. A male with one of these diseases, on the other hand, is X^dY, where X^d is an X chromosome carrying a defective gene which, for example, is responsible for the lack of a blood-clotting factor in haemophiliacs. Thus, the two possible combinations which can be produced by uniting a sperm of a bleeder with an ovum of a normal woman are XX^d girls and XY boys. The boys are normal but the girls are 'carriers' of the disease. The carrier has an X^d chromosome with the defective gene and one normal X chromosome but does not suffer from the disease. Thus, there are no sufferers in this first generation. The important point to understand is that these diseases are recessive. That is, if a normal gene is present then its characteristics are dominant over an abnormal one. This explains why an XX^d woman who has one normal X chromosome is a carrier and does not have the disease. The inheritance of these diseases is illustrated in Fig. 4.2.

If a female carrier of these diseases with a normal husband has children, however, then the possible combinations of genes are: XX, XX^d, XY, and X^dY in equal proportions. Thus there can be normal daughters XX, carrier daughters XX^d, normal sons XY, and male sufferers from the disease X^dY. This means, in other words, that a normal man (XY) with a wife who is a carrier (XX^d) has a 25 per cent chance of having an X^dY son suffering from the disease (that is, half of the sons are affected) and a 25 per cent chance of having an XX^d daughter who is a carrier of the disease (similarly half of the daughters are carriers; this is illustrated in Fig. 4.2(b)).

Other diseases are linked, not with the X or Y chromosomes, but with the 44 other chromosomes in the nucleus of a normal human cell. These can broadly speaking be divided into two classes, the dominant and the recessive diseases. They are referred to as *autosomal recessive* and *autosomal dominant* diseases respectively, for all chromosomes apart from the sex chromosomes are known as autosomes (see Tables 4.3 and 4.4 for lists of such diseases).

Genetic screening and selective abortion 51

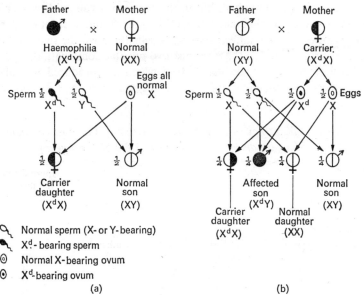

FIG. 4.2. Inheritance of an X-linked recessive disease: (a) affected male mates normal female; (b) normal male mates carrier female.

In a disease caused by a dominant gene, such as Huntington's chorea, dystrophia myotonica, or any of those listed in Table 4.3, each afflicted person has a 50 per cent chance of passing it on to his or her offspring, and both men and women are equally likely to be affected.

Huntington's chorea is a disease whose first signs usually appear when the sufferer is in his thirties or even older. These first signs are an increase in finger twitching, speech disorders which lead to memory lapses, anxiety, and depression; then there is eventually general mental deterioration due to progressive destruction of the brain, which leads to death.

Dystrophia myotonica is a disease which can occur at any age from birth to the mid-teens to the forties. Speech and motor development are retarded and there is also facial weakness and mental retardation.

The way in which these diseases manifest themselves can be seen as follows. If A is the gene that determines whether a person has one of these diseases, then it is found that in the afflicted person there are two different genes, A and a, where a is the normal gene. A is dominant in the *Aa* individual, but in the sperm or ovum, either *A* or *a* will be present in different cells (just like the X and Y chromosomes in sperm). In the unaffected spouse's cells there will be two normal genes *a* and *a*, and so there are just two possible combinations in the fertilized ovum, namely *Aa*, and *aa* which are produced in equal numbers. *Aa* combinations will be affected because *A* dominates

TABLE 4.3

Expected frequency of occurrence in Britain of the most common dominant diseases (assuming 800 000 births a year)

	Approximate number of births a year
Deafness*	80
Blindness*	80
Epiloia (tendency to form certain tumours in skin and other tissues)	40
Marfan's syndrome (abnormalities of connective tissue)	40
Achondroplasia (form of dwarfism)	40
Neurofibromatosis (skin tumours)	40
Dystrophia myotonica (muscle wasting)	40
Huntington's chorea (see p. 51)	80
All others	120
Total	560

* These totals include the effects of several separate diseases, which collectively make up the totals shown. The diseases without an asterisk can be assigned to a single defective gene.

The frequencies quoted, which are based on figures provided by Professor J. H. Edwards, are subject to error because of the difficulties of obtaining reliable statistics for such rare diseases.

over *a*, the normal gene, in causing the disease. The offspring therefore has a 50 per cent chance of being affected (Fig. 4.3). Thus if some tests could be devised to pick out the combination *Aa* early in pregnancy, then, if the parents so desired, these foetuses could be aborted. Eventually if such a programme of selective abortions was pursued vigorously, the occurrence of the disease might diminish substantially. Unfortunately, at present there are no reliable tests available that can be carried out to detect early in pregnancy any diseases caused by dominant genes.

Most diseases caused by single genes are what are called recessive diseases, and by far the greatest success in diagnosis at an early stage in development of the foetus has been obtained with these diseases (see Table 4.4). The diseases which can be detected are those that

Genetic screening and selective abortion

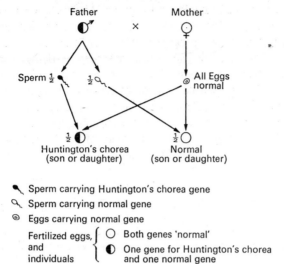

FIG. 4.3. Inheritance of a dominant genetic disease. Huntington's chorea: father affected, mother normal. The results would be the same if an affected female mated with a normal male.

show anomolies when the amniotic fluid and the cells it contains are subjected to biochemical tests. Table 4.4(b) lists 40 diseases which can be detected by biochemical tests. This number, however, is small compared with the nearly 800 known recessive diseases, which in turn are a small proportion of all those which, in principle, can occur. Most of them, fortunately, occur only very rarely. The gene which determines the characteristics of any one of these recessive diseases is much more common than the disease itself. This is because, in contrast to diseases caused by dominant genes, two copies of such a gene must be present before the disease manifests itself—one from the mother and one from the father. Even then the probability of any child of such a union having the disease is only 1 in 4. This can be seen as follows. Suppose the normal dominant gene is B, and the recessive gene causing the disease b. Both husband and wife have the genetic make-up Bb, so that there are four possible, equally likely, combinations of genes in their offspring, namely, BB, Bb, bB, and bb (a B gene or a b gene from either parent). Only a child having the last combination (bb) will have the disease, although children with a Bb combination will be carriers of the disease, just like the female carriers of haemophilia. This is illustrated in Fig. 4.4.

Now a case can be made out for prenatal screening to see whether the child which has been conceived has the genetic make-up *bb*. Any decision to terminate a pregnancy based solely on the fact that both mother and father carried a recessive gene for a particular disease will mean that three normal babies will on average be lost for every affected foetus. It is usually a far easier process to determine the genetic make-up of the parents than that of the unborn child. In principle, the procedure to be adopted in such a case might therefore be first to screen the wife for the presence of the recessive gene. If it is present, the husband would be similarly screened and only if the

TABLE 4.4

(*a*) *Most common recessive diseases in Britain and their expected frequency of occurrence* (*assuming 800 000 births a year*)

	Approximate number of births a year
* Severe mental defects (excluding aminoacidurias)	640
Cystic fibrosis	400
* Deafness (severe, all forms)	400
* Blindness (severe, all forms)	160
* Adrenogenital syndromes	80
Albinism	80
Phenylketonuria (PKU)	80
*† Aminoacidurias (excluding PKU)—treatable	40
*† Aminoacidurias (excluding PKU)—untreatable	40
*§ Mucopolysaccharidoses (all forms)	40
Tay–Sachs disease	8
Galactosaemia	4
Total	1972

(Less than 100 of these diseases could be detected in the uterus in 1973)

* These totals include the effects of several separate diseases, which collectively make up the totals shown. The diseases without an asterisk can be assigned to a single defective gene. Albinism is caused by a defect in one or other of two genes.
† Aminoacidurias are diseases like PKU that are caused by accumulation of an amino acid because of a defect in metabolism. Some 13 different diseases are included under these two headings.
§ These are diseases, like Hurler's syndrome, that are caused by a defect in the metabolism of certain types of sugars, the mucopolysaccharides, whose accumulation in various tissues causes a variety of gross developmental abnormalities.

The frequency estimates given in this table were based mostly on information provided by Professor J. H. Edwards. Because of the low frequencies of these diseases, the numbers given in the table must be taken as approximate.

Genetic screening and selective abortion

TABLE 4.4—contd.

(b) Some recessive diseases detectable before birth

* Adrenogenital syndrome. See p. 57.
Arginosuccinic aciduria. AA.
Citrullinaemia. AA.
Congenital erythropoietic porphyria. Abnormality in formation of an essential non-protein part of the haemoglobin molecule.
Cystinosis. AA.
Fabry's disease. Fat (glycolipid) storage disease due to inability to break down certain types of fats.
Fucosidosis. Fat storage disease.
* Galactosaemia. Inability to use the sugar, galactose; can be countered by a galactose-free diet.
Gaucher's disease. Fat storage disease.
Generalized gangliosidosis. Fat storage disease.
Glycogen-storage-disease-II ⎫
Glycogen-storage-disease-III ⎬ Inability to break down the glycogen sugar complex that is used for the storage of glucose
Glycogen-storage-disease-IV ⎭
Homocystinuria. AA.
Hurler's syndrome. MP.
Hyperammonaemia. Defect in ability to use protein breakdown products as an energy source. Patients dislike protein-containing foods.
Hyperlysinaemia. AA.
I-cell disease. MP and fat storage disease.
Juvenile GM_1 gangliosidosis. Fat storage disease.
Ketotic hyperglycinaemia. AA.
Lysosomal acid phosphatase. Abnormality of nucleic acid breakdown.
Mannosidosis. Inability to use the sugar mannose. Only one case described.
Maple-syrup urine disease. AA.
Metachromatic leukodystrophy. MP.
Methylmalonic aciduria. Defect in ability to use protein breakdown products.
Morquio syndrome. MP.
Niemann-Pick disease. Fat storage disease.
Ornithine-alpha-ketoacid transaminase deficiency. Inability to use protein breakdown products.
Orotic aciduria. Defect in nucleic acid metabolism.
Pompe's disease. A glycogen storage disease.
Pyruvate decarboxylase. Defect in energy metabolism.
Refsum's disease. Fat storage disease.
San Filippo syndrome. MP.
Scheie syndrome. MP.
* Tay–Sachs disease. See p. 62.
Xeroderma pigmentosum. Inability to repair damage caused by ultraviolet light.

* Only the diseases marked with an asterisk are given a frequency in Table 4.4(a). Some of the other diseases, such as the aminoacidurias and mucopolysaccharidoses, however, contribute to the overall frequencies for their disease class, as given in Table 4.4(a). The remaining diseases occur so infrequently (often just a few isolated families have been described) that no reliable figure for their incidence can be given.

A very brief indication of the nature of the diseases has been given except for those discussed in the text. AA denotes aminoaciduria; MP, mucopolysaccharidoses: these diseases are described at the foot of Table 4.4(a).

The fact that many of the diseases have a similar biochemical basis is mainly a reflection of the available techniques. Only those diseases whose basis can be unravelled using available techniques are in this list. This number is steadily increasing.

recessive gene was found to be present in both would it be necessary to consider screening the foetus. At present, because there are virtually no large-scale screening programmes, amniocentesis is rarely considered unless the parents already have an affected child. The extent to which such a practice decreases the number of affected births is discussed later in this chapter.

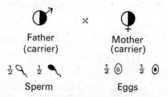

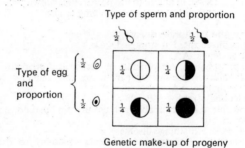

FIG. 4.4. A cross between carriers of the deleterious PKU gene. White areas in the circles represent the normal gene; black areas the PKU gene. (Adapted from L. Cavalli-Sforza.)

Out of 800 000 or so births, about 600 mentally defective children are born in Britain every year as a result of single-gene defects. (Not all the diseases are due to the same defect, but several defects add up to give this total.) About 400 deaf children also are born annually; again this total includes several different types of diseases. The most common disease which can be identified with one particular gene is cystic fibrosis of the pancreas, which is discussed in detail later in this chapter. About 400 such children are born each year in Britain. In

Genetic screening and selective abortion 57

contrast to deaf and mentally defective children, whose life-expectancy is not vastly different from that of normal children, cystic fibrosis sufferers are not, at present, expected to live past the age of 20.

Another common genetic affliction in Britain is blindness; about 160 blind children are born every year. Here again, though, the disease is not identified with one deleterious gene but several diseases contribute to make this total. About 80 children are also born every year suffering from the adrenogenital syndrome; albinism and phenylketonuria account for similar numbers. All three diseases can be identified with single genes.

In children suffering from the adrenogenital syndrome, the adrenal glands are overdeveloped. The result is that an excessive amount of the male sex hormone androgen is secreted. In baby girls this can cause excessive masculinization of the external genitalia. The clitoris may enlarge to become penis-like and the labia may develop into a type of scrotum. In male children the syndrome may manifest itself as the 'Infant Hercules' syndrome, in which the child reaches puberty within a few months of birth.

If children suffering from a malfunction of the adrenal glands are recognized as such early in life, then there is every chance that with treatment they can lead normal lives. If not detected, however, the disease will prove fatal.

Albinism is a disease in which the pigmentation of the individual affected is much lighter than normal. There is, however, no reason why a person suffering from albinism should not lead a more or less normal life. Such people, however, are subject to high risks of blindness and they have to take greater care of their eyesight than normal individuals.

Phenylketonuria is a disease affecting the metabolism and is closely related to albinism (as discussed below), but its effects are much more serious. Children suffering from phenylketonuria will almost invariably be mentally defective unless action is taken soon after birth. The other diseases shown in Table 4.4 occur even more infrequently in Britain.

Although a large number of recessive diseases can be detected within the womb, two of the most common cannot be detected reliably early in pregnancy by commonly available techniques. These are cystic fibrosis, which occurs in about 1 birth in every 2000, and phenylketonuria, which occurs in about 1 birth in 10000, both figures referring to Europeans and caucasian North Americans. Sickle-cell anaemia, which is very common among people of African origin, with an incidence that can be as high as 1 birth in every 40, also cannot be reliably detected early in pregnancy.

Polygenic diseases

Polygenic diseases are caused by the combined action of several genes, but may also be strongly influenced by environmental factors. Two of the most common polygenic diseases are the congenital malformations spina bifida and anencephaly. In Britain about 2000 children suffering from spina bifida and 2000 children suffering from anencephaly are born every year; 4000 other children suffering from various congenital malformations which are due to polygenic disorders are also born annually in Britain.

As these diseases have no simply determined hereditary component there is so far no possibility of predicting before conception that an affected child is likely to be born. There seems in fact, to be an environmental factor involved in, for example, spina bifida and anencephaly, and a mother who has had an affected child has about a 5 per cent chance of having another. More than 90 per cent of affected children are, on the other hand, born to families with no previous record of the diseases.

During 1973 it became possible to detect most cases of both these diseases in the uterus. The test is based on the occurrence of a particular protein (alphafoeto protein) in amniotic fluid. The damaged tissues apparently leak this protein into the fluid, which can then be drawn off during amniocentesis.

In spina bifida the spinal cord and nerves are exposed, and in severely affected cases death may occur at an early age. In many cases, however, operations can be performed to correct these outward defects but many of these operations unavoidably cause severe damage, leaving the child physically sound but mentally retarded. Anencephalic children normally die soon after birth.

Even though a routine test is available to detect anencephaly and spina bifida in the uterus, it is unlikely that it will be universally applied, for the costs and dangers of amniocentesis will outweigh the advantages of screening unless there is a method available for identifying mothers at risk. At the end of 1973 it seemed that it might be possible to detect both diseases simply by testing the blood of the expectant mother. An increased amount of alphafoeto protein in the blood of the mother-to-be seems to be associated with a diseased foetus. The test of the mother's blood is, however, not yet completely reliable: some false positive diagnoses are made. More work is needed to make this test on the blood reliable. With a test of the blood available at a later stage of pregnancy, it should be possible to confirm whether or not the foetus really has anencephaly or spina bifida, without having to sample the amniotic fluid (see below).

If the test of the mother's blood is carried out as a routine investigation, then the cost could come to less than £1 for each test. In Britain the cost would be about £800 000 a year for investigating every pregnant woman. These are very tentative figures based on preliminary results but they give hope that in the near future both anencephaly and spina bifida may be predicted as a matter of routine in plenty of time for the foetus to be aborted.

Amniocentesis

Amniocentesis, the process of obtaining a sample of amniotic fluid from around the foetus, is possible after the eighth week of pregnancy, but is not safe before the 14th week of pregnancy. Most tests are carried out at 16 weeks. It is necessary to wait until the pregnancy is relatively well developed because of the need to obtain about 15-20 millilitres of the fluid to carry out the tests and to ensure that the foetus is not damaged by the procedure for obtaining the fluid sample. At 16 weeks there are usually between 160 and 320 ml of fluid surrounding the foetus with an average of 175 ml, and so a loss of 15-20 ml can be tolerated.

Before a fluid sample is taken the position of the foetus is determined by an ultrasonic technique. In this process sound waves are used to define the foetus's position. This is a relatively new development, which complements X-ray analysis, but is potentially much less dangerous. Such a procedure is almost essential to exclude the possibility of twins and if the operation of obtaining the sample of fluid is to be anything but random.

The fluid sample is best obtained under local anaesthetic by inserting a needle into the patient's abdomen (Fig. 4.5). In this way the possibility of accidentally inducing an abortion is small and there is little risk of introducing infection. The ultrasonic investigation will show the best place to insert the needle.

At the present state of the art, amniocentesis, even though it causes little discomfort to the patient, is carried out only when there are clear indications that the foetus might be affected. The mother, before agreeing to undergo amniocentesis, is often asked to agree in principle that she will be prepared to undergo an abortion if the prognosis is unfavourable, but she should, of course, be left free to make her own decision when the time comes.

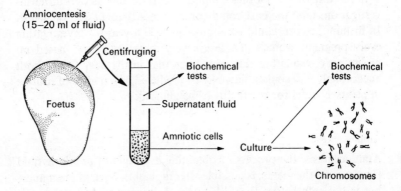

FIG. 4.5. Prenatal detection of inherited disorders by study of cells from the amniotic fluid. Biochemical tests of cultured cells can, for example, reveal enzyme deficiencies; analysis of the chromosomes within the cells can detect Down's syndrome (mongolism).

A risk is attached to the process of amniocentesis. Possible dangers to the foetus include, for example, death from a ruptured placenta, foetal haemorrhage, or premature labour. The available data suggest that the foetus could be lost after about one out of every 25 amniocenteses. For example, of 200 amniocenteses which have been analysed by M. A. Ferguson-Smith and his colleagues at the University of Glasgow, there have been 4 cases of premature labour followed by loss of the baby. It is not, however, clear how many of these were the direct result of amniocentesis, since a small proportion would be lost in any case. At present relatively few obstetricians are practised in amniocentesis, but once expertise is obtained the procedure will probably develop into a normal minor surgical practice.

Most antenatal tests for genetic diseases depend on establishing a culture of the cells of the foetus which have been obtained from the amniotic fluid. The analysis is then carried out in a laboratory which

Genetic screening and selective abortion

has facilities for analysing these cells in tissue culture. The tests which can be carried out are of two types: analysis of the chromosomes within the cells, or biochemical tests of the culture to detect the presence, or absence, of certain chemicals.

Table 4.5 shows some published results of antenatal diagnoses from medical laboratories all over the world. Of 322 pregnancies examined, 41 were affected by some disease and 39 patients, 12·7 per cent of the total, had their pregnancies terminated. But the initial 322 patients were all, to some extent, at risk and the percentage of pregnancies terminated bears no relationship to the percentage which would need to be terminated with a completely random sample of pregnant women. These results show that there are three classes where a high proportion of affected foetuses have been found: first, where the parent has a known chromosome defect; second, where the mother is the carrier of a gene linked to the X chromosome; and third, where both parents are carriers of a recessive gene. Two other classes which are fortunately associated with a lower risk than those mentioned are older mothers, who have a higher risk of having a mongol child, and mothers who have previously had a mongol child. (Women aged 45 or older have a forty times greater chance of having a mongol child than women 20 years younger). In both cases 4 per cent of pregnancies in the admittedly limited sample shown in Table 4.5 were affected.

It is essential to have diagnosis before birth completed as early as possible in pregnancy without risk to mother or foetus. If termination of the pregnancy is then thought necessary and agreed to by the parents, it can be carried out before the foetus becomes 'viable'—although when that is, is in itself an arguable point. In practical terms this means that a definite diagnosis has to be available by the 20th–22nd week of pregnancy, although in Britain the law allows pregnancies to be terminated up to the 28th week. That gives the laboratory 4 to 6 weeks at most to complete their test if the amniocentesis is carried out in the 16th week of the pregnancy. If the expected disease is associated with some defect in the chromosomes, then the results are available on average within 11 to 15 days. But if the test is based on the presence or absence of some enzyme or chemical, where a relatively large number of cells are required for diagnosis, then a longer time must be allowed for enough cells to be grown in the laboratory from the foetal cells obtained from the fluid. This procedure might well take 6 to 8 weeks, which then takes the pregnancy up to 24 weeks.

Even if diagnoses take longer than this and the stage is reached where it is not feasible, or lawful, to terminate a pregnancy then they may still be useful in that, in certain cases, treatment for the baby can start soon after birth. Also, after 28 weeks, an indication that the baby is normal should be a great relief to the mother and father.

TABLE 4.5
Experience of antenatal diagnosis by amniocentesis

Indication for amniocentesis	Total pregnancies examined	Total pregnancies affected	Per cent affected	Total pregnancies terminated
Previous child with Down's syndrome	80	3	3·8	2
Other previous chromosome abnormality	1	0	0	0
Chromosomal translocation	38	12	31·6	12
Maternal mosaicism (cell mixture)	1	0	0	0
Advanced maternal age	119	5	4·2	5
Family history of Down's syndrome	4	0	0	0
Exposure to mutagens	9	0	0	0
Sex-linked diseases	31	12	38·7	11
Recessive autosomal diseases	35	9	25·7	9
Others	4	0	0	0
Total	322	41	12·7	39

Based on data provided by Professor Malcolm Ferguson-Smith summarizing ten independent studies reported between 1967 and 1972 from North America and Europe.

A great deal of research is at present directed towards ensuring that diagnosis can be completed in an even shorter time. New techniques of growing cells in culture are being studied and improved biochemical methods are being developed so that a smaller sample is needed for a successful diagnosis.

Tay–Sachs disease

Because the necessary tests have not yet been developed, screening programmes for preventing the birth of children suffering from any of the most common genetically determined diseases are not available. But a programme for screening for Tay–Sachs disease, which has a low frequency in the general population but a relatively high fre-

Genetic screening and selective abortion 63

quency in certain Jewish groups has been operating successfully in the Baltimore–Washington area, Toronto, and Montreal. In 1973 a screening programme for this disease was set up in London.

Tay–Sachs disease is a recessive disease and is especially prevalent among Ashkenazi Jews (see Table 4.4). It is invariably fatal by the age of 4 and there is no known cure. In Britain the overall incidence of this disease is about 8 births every year. But the affected gene is ten times more likely to be present in Ashkenazi Jews than it is in other people and so the disease itself is a hundred times as frequent in these Jewish populations as elsewhere.

The screening programme set up in London at the Hospital for Sick Children in Great Ormond Street was described in *The Lancet* of 17 November 1973. The programme is voluntary and the hospital is prepared to accept blood from Jews or others who consider themselves to be in particular danger of being carriers of the Tay–Sachs gene. For reasons of efficiency, however, the hospital prefers to take blood in mass sessions, at colleges or synagogues for example. About 5 ml of blood has to be taken and it is tested for the presence of an enzyme, hexosaminidase A. Carriers of the Tay–Sachs gene have an appreciably lower amount of this enzyme in their serum than normal people.

The results of screening 7000 Jews in North America for the presence of the affected gene resulted in 300 carriers being identified, but in only 11 cases were carriers married to each other. That is, 11 marriages at risk were identified. In the London programme 175 Jews had been screened up to November 1973 and 6 carriers identified; 14 other carriers were, however, identified among close relatives of affected children.

Once a marriage at risk is identified, prenatal diagnosis is carried out for each pregnancy to determine whether or not the foetus is affected. In the programme at the Great Ormond Street Hospital four separate analyses of the amniotic fluid are carried out. Three of these analyses entail growing the cells in a culture for some days. The fourth test merely involves a direct quantitative test of the fluid for the presence of the enzyme. No discrepancy was observed between the results of the four tests and the procedure eventually adopted was to take direct analysis as definitive and recommend action based on this result. In this way it is known whether or not the unborn child suffers from Tay–Sachs disease very soon after the amniocentesis is carried out. Abortion, if the mother-to-be decides she wants it, can then be carried out at an earlier stage than would be possible if the results of cell culture had first to be obtained.

Because the screening in London had been going on for only a relatively short time, only 10 pregnancies had been monitored for Tay-Sachs disease up to November 1973. Of these, 3 foetuses were predicted to be affected, all of which were aborted. They were shown to have the disease by studies after abortion. Of the 7 foetuses which were predicted to be normal, 4 were born after a full pregnancy and were all normal. In November 1973 one was still not born, one had been spontaneously aborted after a car crash, and the seventh had been deliberately aborted. All the diagnoses were in fact proved correct.

The doctors who run this screening programme stress that they present the facts of the disease to the patient who comes seeking a test for the presence of a carrier gene. It would seem at first sight that there would be a need to screen only in antenatal clinics, but there are disadvantages in this approach. Tests on the serum of pregnant women can lead to false results though a test on the hexosaminidase A content of the white blood cells can give unambiguous results. This, however, is not the principal reason why screening is not carried out during pregnancy. In the *Lancet* article it is pointed out that if a pregnant woman finds out that she is a carrier, even if her husband is not, then this is likely to distress her. The rationale behind the programme is summed up as follows: 'We prefer to screen both partners at the same time, or to test before marriage, having in mind planned parenthood of which abortion is only a part.'

Cystic fibrosis

The most common disorder in Britain and most other western countries caused by a recessive gene is cystic fibrosis of the pancreas. About 1 child in every 2000 is affected, which implies that 1 person in every 20 or 25 is a carrier of the altered gene. This disease cannot yet be detected by amniocentesis although there is a possibility that it will in the near future.

Cystic fibrosis is a severe disease which in most cases leads to death before the child reaches adulthood. In about a tenth of all cases the intestines are acutely obstructed shortly after birth and a hazardous surgical operation is needed to clear the obstruction. If the baby survives this operation, then it is likely to progress as well as other children with the disease. In most children, however, the onset of the disease is more insidious, with a persistent cough and recurring chest infection which may include severe bouts of pneumonia. Other children fail to absorb food from the digestive system and they are therefore slow to gain weight. Most sufferers show both symptoms, though one or the other becomes evident first.

All the effects mentioned are the direct consequence of ducts in the body being blocked by mucus. In the lungs, small air passages are blocked, and infection spreads and develops behind the blockage. In the gut the duct of the pancreas is blocked and as a result the enzymes which cause food to be digested (trypsin and pepsin), which are secreted by the pancreas, build up pressure in the gland and escape into its own tissues. There they destroy the gland, which eventually becomes replaced by a mass of fibrous tissue. The end-result is that the enzymes do not reach the food and protein passes through the body undigested. Similar blockage of bile ducts leads to fibrosis of the liver, but this is less severe in its effects. In boys the sperm ducts are blocked and they eventually become obliterated, resulting in sterility.

The failure of the body to absorb food can be compensated by giving pancreatic extract by mouth. The chest infections can to some extent be prevented by a combination of physiotherapy and preventive antibiotics, each infection that occurs being vigorously treated. In the past few years this line of treatment of cystic fibrotic children has steadily improved, so that their life-expectancy has gradually risen, but few are still able to reach adulthood. All patients eventually die of chest infections or their consequences and will continue to do so unless a more specific and adequate form of treatment is developed.

Cystic fibrosis can now be diagnosed, in principle, by a standard test after birth which makes use of a strange observation. That is that the sweat of children suffering from this disease contains an excessive amount of common salt (sodium chloride). Similar increases in salt have been found in saliva and other body secretions. Available evidence suggests that at least some of the observed effects are due to a factor circulating in the blood. The current belief is that this factor in the blood leads to a change in the membranes of the body cells thus making them more porous.

As yet, no cheap and reliable test has been developed which gives a minimum of false results. A false positive result can be accommodated within a screening programme in that attention will be directed at the child so identified; eventually, it is probable that it will be realized that the original result was false. But with an incidence of 1 in 2000 for cystic fibrosis it is not feasible to test all children twice to determine if a false negative has been recorded.

The first test for cystic fibrosis is based on an analysis of sweat, as previously mentioned. The standard test using this method consists of collecting sweat over an area of skin, weighing it, and then determining the concentration of sodium or chloride in the sweat. This test is reliable when the child is over 3 months old provided enough sweat can be obtained. The test is laborious to perform, however, and most young babies are not prone to sweat profusely. Also, perhaps of greatest importance, there are too many false negative results when the test is carried out on babies under 3 months old.

Another test is based on an analysis of the baby's first bowel movement after birth (the meconium). This usually contains little undigested protein but appreciable amounts are present in cystic fibrotic children. Methods based on detecting the protein in the meconium seem to hold out some hope for the future. Various chemical methods of analysing the meconium are now under trial. They have the advantage of being cheap and simple to perform, but unfortunately they give some false positive readings for premature babies. It is too early to say how often a false positive will be obtained for older babies. Of greater importance, it is not known what the false negative rate will be with this method.

If prenatal diagnosis is not yet possible and postnatal tests so unreliable, a way of controlling the disease must be looked for by going back one step further and searching for carriers of the recessive cystic fibrosis gene, which is probably present in one person in each 20 or 25 of the population. Screening infants for cystic fibrosis makes it possible to start treatment at the earliest moment, although early detection does not reduce the number of affected children. But if the carriers of the gene could be recognized through a screening procedure carried out before they become parents, then two affected people marrying could at least be warned of the problems of having children.

The one test currently available for detecting carriers of the cystic fibrosis gene is the 'cilia' test. This test is based on observing what happens when the serum of patients is mixed with the cilia (hairs) of fresh-water mussel gills or oyster gills. If the patient is a carrier then

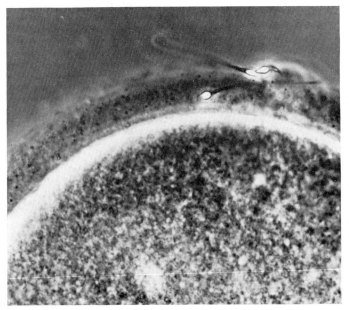

1b. Sperm about to penetrate the egg

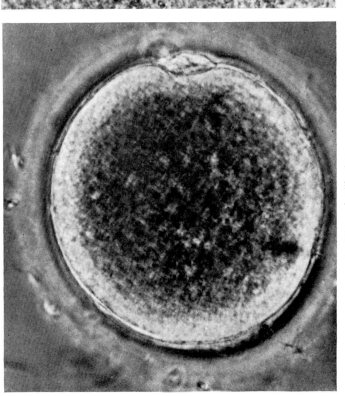

1a. The unfertilized egg

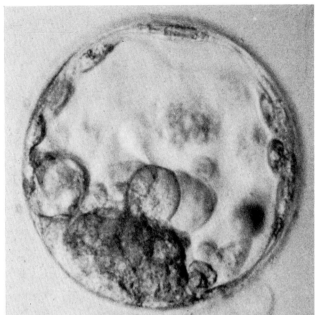

2a. The four-cell stage—two cleavages after fertilization

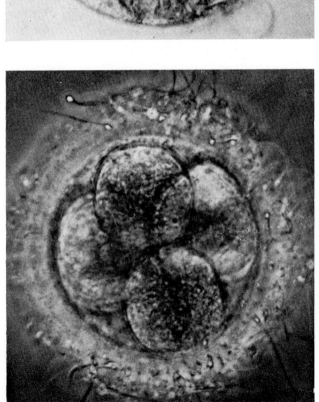

2b. The human embryo 5 to 6 days after fertilization—now called a blastocyst. Only the inner cells actually form the future foetus. The remaining cells which form the outer layer of the spherical blastocyst give rise to the placenta.

the beats of the cilia are inhibited. This test has proved a useful research tool in the absence of a better method, although it has a high false positive rate of about 1 test in every 10. Unfortunately, it does not yet give positive results in amniotic fluid, cord blood, or often in young infants. It is now known, however, that growing a tissue culture from carriers of the disease releases into the tissue culture medium the factor which inhibits the beat of the cilia. If this factor is to be found in amniotic fluid cells it could therefore in principle be detected from a culture of those cells. Because of the high false positive rate the cilia test is, however, not yet even an acceptable test for routine clinical diagnosis for screening for gene carriers. It is used on an experimental basis to detect carriers of the disease, although sufferers of the disease show a similar response. It may therefore be difficult to distinguish in the uterus between carriers of the gene and the actual sufferers. If it proves possible to refine the test so that the carriers and the sufferers can be distinguished, possibly by some kind of quantitative estimation of the inhibition of the cilia, then the chances of eventual successful diagnosis in the uterus will certainly be increased.

Phenylketonuria

The detection and treatment of cystic fibrosis leaves much to be desired, but more success has been obtained in the detection and treatment of another genetically recessive disease, phenylketonuria, which like cystic fibrosis cannot yet be detected before birth.

Phenylketonuria is a disease caused by the body's inability to metabolize the amino acid phenylalanine, which is found in protein. Failure to metabolize phenylalanine can lead to several hereditary diseases: alkaptonuria, in which urine turns black when it is exposed to air; tyrosinosis, where the body accumulates tyrosine, another of the amino acids present in protein; albinism, where pigment is not present in the skin and hair; and phenylketonuria which is a form of idiocy. Fig. 4.6. shows the way in which phenylalanine is normally metabolized in the body. At least five genes take part in the process and the four diseases mentioned are caused by the respective failure of these genes to operate.

If gene 3 is defective, homogentistic acid is not broken down into water and carbon dioxide and so is excreted in the urine. This causes the condition known as alkaptonuria where the urine of sufferers turns black on exposure to light. These people also tend to develop

68   *Genetic screening and selective abortion*

arthritis. If gene 2 is defective, then one of the chief ways in which tyrosine can metabolize is blocked. The result is that an excess of an immediate product of tyrosine is formed, which is excreted in the urine. Fortunately, there are no further ill effects.

But if genes 1, 4, or 5 are defective, then the consequences are more serious. Genes 4 and 5 are required to change tyrosine into melanin, which is a pigment, and persons so afflicted cannot produce this

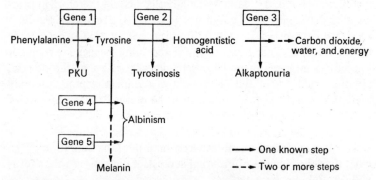

FIG. 4.6. Simplified diagram of the way in which phenylalanine is metabolized in the body. Phenylalanine and tyrosine are amino acids, two of the twenty basic subunits of proteins. Homogentistic acid is an intermediate product in the metabolism of tyrosine. Genes 1, 2, 3, 4, and 5 act at the points shown, and when they are defective result in the corresponding diseases: gene 1, PKU (accumulation of phenylalanine); gene 2, tyrosinosis (accumulation of tyrosine); gene 3, alkaptonuria (accumulation of homogentistic acid); genes 4 and 5, albinism (non-production of melanin).

pigment in their skin, hair, or eyes. These people are albinos and inevitably have weak eyes.

By far the most serious effects occur if gene 1 is defective. Phenylalanine cannot then be converted directly into tyrosine, and this defect in the body's metabolism causes severe mental defects. If left unattended, a child who is born with this deficiency will usually end up severely mentally retarded. This is the defect known as phenylketonuria, or PKU for short.

PKU does not occur as frequently as cystic fibrosis but is still common enough to cause great concern. In the United States it is estimated to occur once in every 14 000 births. In the Netherlands the figure is quoted as 1 in 26 000 births. By contrast, in West Germany it is diagnosed once in every 6000 births. In Britain the incidence varies from area to area with an incidence of 1 in 21 000 reported for north London, 1 in 5900 in Glasgow, and 1 in 6300 in Manchester.

Genetic screening and selective abortion

PKU can usually be detected immediately after birth, and screening for the disease is undertaken in all parts of Britain and in most European countries. PKU screening in Britain was prompted by the Department of Health and Social Security in 1969 and by 1973 more than 85 per cent of babies born each year were screened for the disease.

In contrast to cystic fibrosis, the prospects for children suffering from PKU are good if they are detected early in life. When they have been identified as suffering from PKU, children are put on a special diet that is low in phenylalanine. Most protein contains between 5 and 10 per cent of phenylalanine and so these children must be fed with specially prepared protein that is low in this amino acid. Appropriate fats, carbohydrates, minerals, and vitamins are added to the protein. The body usually requires between 35 and 50 milligrams of phenylalanine per kilogram of body weight each day in the first year of life, with a slow reduction to 20–25 mg per kilogram of body weight as the child grows older. It is advisable, to prevent the infant's brain being damaged by high phenylalanine levels, for treatment to start within the first month of life, and the diet must be carefully maintained for at least the first 5 or 6 years of life. Thus, for the treatment to be successful, the child has to be in contact with an experienced dietician who has access to a laboratory for monitoring the child's blood and urine. This is essential to prevent over-treatment with the possibility of protein malnutrition, anaemia, and loss of weight, any of which can cause the child to die.

This form of treatment for phenylketonuria sufferers was proposed by the British Medical Research Council in the middle 1960s but it came in for some criticism. Now that many children have been on this treatment for several years, a detailed analysis and evaluation have been carried out. For 102 infants treated before they were 3 months old, the average I.Q. was 94 after a few years, with individual values ranging from 73 to 120, well within the 'normal' range. The skin lesions, which are another symptom of the disease, cleared, and the fits which such children usually suffer from either did not develop or ceased after treatment. Their hair also became darker, demontrating that more melanin was being created in the body (see Fig. 4.6).

The treatment of PKU and postnatal screening for the disease is a clear-cut example of where, in terms of costs, the benefits of screening seem obvious. This can be seen in figures taken from the PKU screening unit in Manchester for 1972. The unit tests 70 000 babies a year, and among these about 11 cases of children suffering from PKU are found. The budget for the screening programme is £7400 a year and to maintain each child on a diet for 6 years and to pay the dietician costs a total of £38 500 a year. The cost of detecting and treating each case is thus about £4200. But what if the disease had not been detected? Then the child would inevitably have landed up in an institution. If the life-expectancy of each sufferer is, say, 40 years then the costs of having to maintain the patient for at least 30 years would have to be borne. Conservatively this comes out at £800 a year, so that over 30 years it would be £24 000, which is considerably more than the £4200 that the treatment would cost. In addition there is no loss of earning capacity and the family will not be distressed by having a mentally retarded child. It is, of course, likely that putting a child on a special diet may place some strain on the family, but this is a small price to pay for preventing the mental retardation.

The test for PKU is cheap and easy to perform, and in Britain it is recommended that it be carried out between the sixth and fourteenth day after birth. The method most frequently used is the Guthrie blood test, which is based on a growth test with special strains of bacteria. It enables the levels of phenylalanine in the blood to be estimated. Several spots of blood from the infant are collected, usually from a heel-prick, on to specially absorbent filter paper. In the laboratory, a small disc is punched out of the blood-impregnated filter paper and about a hundred such discs are placed on a special plate containing bacteria and an inhibitory substance. Phenylalanine acts as an antagonist to the inhibitor, and so after incubation a growth will be observed around those discs which contain phenylalanine. The amount present can be estimated by comparing the size of the growth circle with growths obtained using known amounts of phenylalanine (Pl. 3(*b*)).

The cost of this test is about 15p for each child in Britain, while in the United States the cost is estimated at $1.25 per child. The cost includes materials for taking blood, collection and delivery to the laboratory, postage, and laboratory expenses, including the salaries of technicians and secretaries. But the cost does not include the doctor's time or that of the health visitor. It is, however, important to point out that it is hard to decide just how to include overheads in these cost estimates. Thus, including a reasonable fraction of the hospital overhead might considerably inflate the screening costs,

Genetic screening and selective abortion

while ignoring overheads would minimize the estimated cost of hospitalizing a few retarded children. Even allowing, however, for these error margins, the straight economic benefits of PKU screening, at least in Britain, seem to be clear cut.

The PKU screening programme satisfies all the criteria needed for screening programmes. The tests are technically reliable with a very low proportion of false results, either positive or negative. Cases diagnosed as positive are confirmed and investigated for any unusual clinical and biochemical problems. The cost is low and the treatment is beneficial to the patient, his family, and to third parties. The treatment is also safe.

There is no fear that cystic fibrosis sufferers will become parents, for most of them die before adulthood and the males, whose sperm ducts are usually obliterated at an early age, are in any case sterile. But the situation is different with PKU sufferers. The treatment is good and reproduction is possible. There are also some untreated PKU sufferers in the population with normal or below-normal intelligence. The few widespread surveys which have been carried out in the United States indicate, however, that the incidence is less than 1 in 100 000.

What are the dangers of women with PKU having children? If a PKU mother marries a man who is not a carrier of the deleterious gene for PKU, then all their children will be carriers, and none should have the disease. Normally, carriers of the gene will not need the attention which is accorded to children who suffer from the disease. But when the mother is a sufferer, it is virtually certain that all her children will be mentally retarded unless she is on a diet during her pregnancy. The excess phenylalanine in the mother is carried over to the foetus and causes the disease even if the foetus has a normal gene and so is only a carrier. There is, however, no further danger to the child once it is born.

There is no way of telling without testing every pregnant woman whether she is a PKU sufferer. The incidence of PKU sufferers in the population seems, however, not to be high enough to warrant a screening programme designed to detect these women. Now that screening programmes at birth have been instituted, it will on the other hand become essential to keep track of any woman who has been identified as a PKU sufferer. The choice facing a PKU mother will then be either abortion or the special diet to try to prevent the child from being mentally retarded.

Sickle-cell anaemia

Sickle-cell anaemia is a disease in which, in certain circumstances, the blood's ability to carry oxygen is impaired. The oxygen in the blood is carried by haemoglobin, which is the active ingredient in the red blood cells. In sufferers from sickle-cell disease there is an unusual form of haemoglobin, which is formed by a corresponding abnormal gene. This abnormal haemoglobin behaves in a way similar to normal haemoglobin when there is an abundant supply of oxygen, but when sickle cell haemoglobin loses oxygen, as it does in the body when the haemoglobin delivers oxygen to the tissues, the abnormal behaviour appears.

Red blood cells are normally disc-shaped, but sickle cells when they lose oxygen become distorted. The affected haemoglobin molecules attach themselves to other affected haemoglobin molecules to form rigid rods, which distort the red blood cells. The result is that the red blood cells adopt the form of a sickle—hence the name for the disease (Pl. 4(a)). The sickle-shaped cells tend to clog the blood-vessels of the body, especially the capillaries, and the result is that the sufferer becomes anaemic.

Many sufferers die in childhood though an increasing number survive until their thirties or even longer. Sufferers from the disease do not have an easy life, for the symptoms include arthritis, leg ulcers, and attacks of acute pain.

This disease was first described in the early part of this century in North America. Since then it has been realized that it is essentially a disease which affects blacks, although it has occasionally been observed in other races. The incidence of the trait, that is the incidence of people with one affected gene who do not suffer from the disease but are carriers, varies across the world. About 1 in every 12 North American blacks carries the gene. The incidence of the disease itself is, of course, much lower and, again in North America, on average between 3 and 13 out of every 1000 blacks examined have the disease, although its incidence can be as high as 25 per thousand in some parts of Africa. The incidence of gene carriers in Central and South America and the West Indies is similar to that found in North America. In Africa there is a higher incidence of gene carriers, although the frequency varies from tribe to tribe. Some tribes are reported to have 46 per cent of their members as carriers, while others only have 19 per cent or less. The gene is found to be prevalent in a broad band which extends across the middle third of the African continent, and it is rare in South Africa.

A feature of the disease in Africa is that there are many more carriers of the trait, but relatively speaking many fewer sufferers, than are found in North America. The ratio of the trait to sufferers of the disease in Africa has been estimated to be about 1000 to 1, whereas in North America the ratio is between 6 to 1 and 30 to 1.

The pattern of occurrence of sickle-cell anaemia and its associated trait is, at first sight, difficult to understand. Sufferers from the disease, that is those people who inherit one sickle-cell gene from their father and one from their mother, generally speaking do not reproduce. Therefore, unless the carriers of the disease are particularly fertile (over and above the normal for other blacks), there would be no reason for the gene to be so common in Africa. There would be a high incidence of carriers if there were a high chance of the normal gene mutating or changing into the sickle-cell gene, but there is no evidence for this. So there must be a selection process of some kind which favours carriers of the sickle-cell gene in Africa.

Children in the area in Africa where the trait is most common are exposed to malaria nearly all the time. The result is that they are repeatedly infected during their childhood, and it has been shown that children who carry the trait are relatively immune to malaria. Children without the trait would thus be more likely to die from malaria than children with the trait. The result is a high incidence of people with the trait.

What has happened to the blacks in North America where malaria is not prevalent? In all probability when blacks were first brought to the United States as slaves some three hundred years ago there was a high incidence of the trait among them. In the United States, however, there was nothing to give children born with the trait such a huge advantage over children born without it. There is no definitive evidence, either, that carriers of the trait are at a disadvantage compared to people with normal genes. The decrease in the frequency of the gene in North America must have resulted from mixed marriages and from the fact that sufferers of the disease normally fail to reproduce.

There is as yet no way of detecting sickle-cell anaemia in the uterus, though it might be possible in the near future to obtain a blood sample from a foetus to test the foetal haemoglobin. There seems, however, to be no great hope at present that this can be done early enough for the child to be aborted. If there were some other way of determining the foetus's blood characteristics, for example, from the mother's blood, then the problem would be solved.

Carriers of the disease can be identified quickly and cheaply, merely by a simple analysis of blood. There might appear to be many advantages in identifying carriers who are at risk with a view to discouraging marriages between carriers, but it is precisely such a programme which ran into trouble in the United States in the early 1970s.

Some state legislatures tried to impose screening programmes on all blacks and with the obvious racial connotations there were protests about discrimination. But it is not only the blacks who are particularly prone to a common genetic disease. The Ashkenazi Jews are prone to Tay–Sachs disease, for example, and screening programmes chiefly restricted to Jews, which are designed to detect and prevent the birth of affected children, seem to be successful both in Britain and North America.

Sickle-cell anaemia has unfortunately received a great deal of adverse publicity in the United States with laws being passed which do not distinguish between the carrier state and the disease itself. Even insurance premiums are sometimes higher both for carriers and sufferers, though there is no medical evidence to support the implication that the carriers are a significantly poorer risk.

Screening programmes

Genetic screening programmes differ in one fundamental respect from the traditional framework of medicine. Within this framework a patient goes to a doctor, who is expected to act in the best interest of the patient. The doctor who has been approached will either consider himself competent to advise the patient or he will refer him to someone with an appropriate specialized knowledge. Eventually the patient consults a medical practitioner who is competent to give advice, which the patient may or may not take.

Doctors are not infallible and incorrect diagnoses, inappropriate referrals, negligence, and other problems will occasionally arise which sometimes will be evident to the patient. Fortunately, such errors can usually be rectified. As the system is essentially voluntary, persons unwilling to refer their symptoms to a medical practitioner are not, in general, forced to do so.

Genetic screening and selective abortion

Genetic screening programmes, on the other hand, work to a different set of principles. Here, the doctor, or some representative designated by a government department, goes to the patient to make various tests and so the usual order of proceedings is reversed.

Parallels to this, of course, are vaccination programmes designed to prevent the occurrence of various diseases of childhood. Here also the usual order of proceedings is reversed and it is the patient who is approached. Such preventative vaccination programmes are a general feature of public health policy in many countries. There is an analogy in water fluoridation, which is designed to improve the dental health of the population. Fluoridation has, however, run into difficulties. Various groups have protested that it is an infringement of the rights of individuals and that the fluoride has harmful effects which tend to outweigh its advantages. Genetic screening programmes and vaccination are different from fluoridation in that each individual is treated separately, although vaccination has at various times, been compulsory.

An important aspect of screening programmes (not necessarily genetic) which is not fully appreciated is that screening for a particular disease is of direct benefit to other people as well as to the person being investigated. Direct benefit to the patient occurs when a person's blood group is established before minor surgery or childbirth. Chest X-rays to search for an early indication of tuberculosis are also of benefit to the patient. Prenatal screening to detect abnormal foetuses is, on the other hand, of greatest benefit to the parents and to other relatives of the unborn foetus. A child born with a severe handicap could be a burden on the family and could deeply affect the lives of existing brothers and sisters. Screening women for rhesus antibodies and syphilis with a view to treatment has also benefited relatives. On a wider front, the screening of school teachers for tuberculosis and of airline pilots for colour-blindness is of direct benefit to third parties as well as to the persons concerned.

A vital difference between screening and conventional medical practice is that the medical practitioner or other person administering the test usually has no expertise in the disease being screened. This is especially so when it is rare, or when the interpretation of the test is complicated as, for example, when cervical smears are analysed. In fact it is possible that the patient exposed to screening, or indeed to propaganda for screening, will not be able to obtain much informed or constructive advice from the person carrying out the test.

It is unlikely that any screening programme will be embarked upon unless something constructive can be achieved with the knowledge gained. The state of medical knowledge is such that there is no benefit to be obtained from the early diagnosis of most genetic diseases. In these and other cases, however, there would be a great deal to gain from a screening programme which would identify the carriers of the disease so that marriages at risk could be identified, or indeed so that marriages between carriers of the same diseases could be discouraged. In view of these difficulties, it is essential that the standards of screening programmes should be high and that the benefits of any recommended treatment be clear.

Although screening for most genetic diseases is not possible, it is nevertheless worth while considering the implications of screening programmes for common and rare genetic diseases in case the techniques for these tests should become available in the future. In particular, it is essential to know the extent to which these diseases can be prevented or indeed eradicated if necessary screening is carried out. And it must be asked whether it is wise to eradicate a gene which causes a disease by such methods.

Such discussion must first of all be centred upon the type of disease being investigated and whether it is caused by a dominant gene or whether it is the result of a recessive gene where two copies, one from the husband and one from the wife, are necessary before the disease manifests itself in any children of the marriage.

If the aim of a screening programme is to prevent the birth of children suffering from a particular disease, then it is best for the carriers of the disease to be identified before they have children. In principle, of course, it is necessary to screen only the foetuses, but this would necessitate a greater effort and be more difficult, even if the techniques for detecting these diseases in the uterus were currently available. A much easier, safer, and less traumatic experience for the couples concerned would be to test either all husbands or all wives before they have children for the presence of deleterious genes.

Table 4.6 shows, for example, the number of women who would have to be screened for two common recessive diseases, cystic fibrosis and phenylketonuria, in order to identify one marriage where there is 1 in 4 chance of a child being afflicted by the particular disease. Once this identification was made, it would then be necessary only to undertake prenatal diagnosis in the identified cases—always assuming of course that reliable tests for detecting the disease in the uterus were available. In Britain there would be a need to study closely every year 1512 pregnancies for cystic fibrosis and 394 for PKU. If the foetuses which had two copies of the deleterious gene could be identified early enough, then 378 and 99 abortions respectively (25 per cent of the number screened) would ensure that no children suffering from these diseases would be born.

Once the matings at risk are identified, the children of such marriages can be investigated to see whether they are also carriers of the gene. In due course, there would probably be no need for mass screening of the population, for all prospective mothers and fathers would know whether they were carriers of the gene. This, of course, ignores spontaneous changes of genes, or mutations, which could produce a carrier of a cystic fibrosis gene from normal parents. These would, however, be a very small proportion of the total number of carriers and can probably be ignored.

In the event of two carriers marrying it would then be necessary to carry out investigations of the foetus during pregnancy. Children with parents who are both carriers of the same recessive gene have a 25 per cent chance of not having that gene in their make-up, a 50 per

TABLE 4.6

(a) *Number of pregnancies at risk every year in Britain that could be identified by screening wives for cystic fibrosis and PKU*

	Cystic fibrosis	PKU
(a) Number of wives screened	529	2025
(b) Number of carriers identified from wives shown in (a)	23	45
(c) Number of husbands screened	23	45
(d) Number of husbands who are carriers and therefore number of marriages at risk identified	1	1
(e) Number of mothers at risk of having affected children in Britain (800000 births a year)	1512	394
(f) Abortions required (i.e., 1 out of 4 of the identified pregnancies at risk)	378	99

This table shows for cystic fibrosis that for every marriage at risk which is identified, 529 wives and 23 husbands would have to be screened; 552 screenings in all. The total number of screenings in the population will be a little more than the number of births per year (all mothers but only husbands of carriers). The expected number of amniocenteses is 1512, of which, on average, one-quarter will be affected. On average, for every affected foetus that is aborted, 4 amniocenteses and $4 \times 552 = 2208$ screening tests will have to be carried out. The average number of screenings for PKU is $2070 \times 4 = 8280$ per abortion.

TABLE 4.6—contd.

(b) *Effects which screening the second pregnancy will have on the numbers of affected children born with cystic fibrosis in Britain*

Pregnancy	Number of pregnancies at risk a year in Britain	Number of normal children born	Number of affected children born
First	1512	1134	378§
Second	1134*	850*	284§
	378†	378†	
Total	3024	2362	662
Total without screening	3024	2268	756§
	(4032)‡	(3024)‡	(1008)‡

* These 1134 mothers, unaware that they are at risk, will have a three out of four chance of producing a normal child, the same chances as the 1512 mothers had for the first pregnancy.

† These 378 mothers, aware of the chances of producing an abnormal child because they already have one affected child, will undergo amniocentesis followed by abortion of the foetus if affected. Only when amniocentesis indicates that the child is normal will the foetus be allowed to develop to full term. This might entail several pregnancies and abortions.

‡ If the parents continued to have children until two normal ones were born without screening then there would be 4032 pregnancies at risk and 1008 affected children would be born.

§ These numbers are just one-quarter of the corresponding numbers in the first column.

cent chance of being carriers, and a 25 per cent chance of having two copies of the bad gene and thus suffering from the disease. If the sufferers from the disease have been identified during pregnancy and aborted, then two out of three of the children born will therefore be carriers.

Such screening programmes as this would not appear too impractical provided the costs were reasonable and the tests were available. Mass screening, however, would be economically sensible only for the most common diseases and for those where the children are expected to survive for a long time—several years—and where the disease is accompanied by a great deal of suffering. No compelling case can be made out for mass screening for a disease which is likely to kill a child within a short time of birth. Here the costs would completely outweigh the benefits, although no analysis can correctly assess the suffering of the parents of such a child.

Selective abortion would be the end-product of these screening programmes, but such programmes will do nothing to decrease the frequency of occurrence in the population of any of the diseases where the afflicted child now either dies before reaching the reproductive age or where the child is unable to reproduce.

Mass screening of the population and selective abortion would ensure that no child afflicted with any of these common diseases would be born. But the costs involved, once the necessary prenatal tests are developed, might turn out to be too high to justify such an approach. A selective approach might then be taken in which the only people to be screened would be those who had already had an affected child.

In such a scheme there is no way which could prevent the first child born from being affected. It can prevent or deter only those parents whose first child was affected from having a second affected child. From Table 4.6 it can be seen that if the families decide to have two children only (whether they are defective or not), then without any screening at all 756 children born out of 3024 would suffer from cystic fibrosis. But with screening followed by abortion after a first birth, the number of affected children would fall to 662, a decrease of 12·5 per cent in the number of affected children born. Of course, if the parents with a defective child wanted to have two normal children this would be possible by monitoring all further pregnancies. But if there were no monitoring of the pregnancies and all the families carried on having children until each family had two healthy ones then to produce 3024 healthy children 4032 would be born, 1008 being defective. In such extremes, screening followed by selective abortion will decrease the number of cystic fibrotic children born from 1008 to 662, a 34 per cent reduction. In practice the reduction would be somewhere between 12·5 and 34 per cent, for it seems likely that parents would decide not to have more children after the birth of a second cystic fibrotic child. The same percentage decrease would apply to other diseases caused by recessive genes.

As about 380 cystic fibrotic children are born in Britain every year without any prenatal screening, such a screening programme after the birth of the first such affected child in the family would decrease this number to anything between 250 and 330. Similarly the number of children suffering from PKU would be reduced only from the present 100 or so a year to between 67 and 88.

The effects of a number of diseases are not observed until the child is several years old. A screening programme after the birth of the first affected child would consequently not be effective for these diseases, because a second child would be likely to have been born before the disease had been diagnosed in the first.

If tests could be developed to detect in the uterus diseases caused by dominant genes and a vigorous programme of abortion were pursued, then the incidence of such diseases could be drastically reduced. They could not be completely eliminated, for natural changes, or mutations, of the genes are always occurring and a normal gene can be changed into an abnormal one. But by vigorously pursuing such abortion programmes the incidence of such a disease could probably be reduced to 1 per cent of its present intensity.

Screening, followed by abortion when the tests show that the mother-to-be is carrying a child suffering from a genetic disease caused by a dominant gene, will reduce the incidence of the disease to that caused by natural mutations within one generation. No such reduction is possible, however, with diseases caused by recessive genes. It is also far from clear whether the complete elimination of the harmful gene is a good thing. This would seem to be a contradiction in terms. Why should not all efforts be made to eliminate the genes which cause cystic fibrosis so that expensive screening programmes with abortions every generation will not be necessary? First of all, many more abortions than would be necessary to eliminate all the sufferers of a disease caused by recessive genes would be needed to prevent carriers of the disease being born. Not only would this mean aborting three out of four (instead of one out of four) of the pregnancies of all the identified matings at risk, but the pregnancies of couples where only one of the partners was a carrier would also have to be monitored so that the two children in four of such a union which would on average be a carrier could be prevented from being born. For cystic fibrosis, there would be a need in Britain to terminate about 35000 pregnancies (out of 800000) annually (that is half of all the pregnancies where either the wife or the husband was a carrier, estimated on the basis of one person in every 23 being a carrier).

There is, however, a more fundamental reason why such a vast screening and abortion programme might not be beneficial. This is that there is some evidence that carriers of some recessive deleterious genes might be able to survive better than non-carriers. The argument is that this might to some extent be a compensation for the fact that sufferers of the disease are at a disadvantage. It has already been mentioned that carriers of the sickle-cell gene are known to be more

Genetic screening and selective abortion 81

resistant to the effects of malaria than people without this gene. The particular advantages, if any, which carriers of cystic fibrosis might have over people who do not possess the deleterious gene, are not known at present, but if there are any then it is not obviously undesirable to have such a trait in humans. There is, for example, some preliminary evidence, from a comparison of 250 families of children with cystic fibrosis with 250 families not prone to the disease, that the families with cystic fibrosis suffered from a significantly lower incidence of tuberculosis. There is no proof, however, that this is connected with the 'bad' gene. It seems likely that every person, on average, carries about three bad genes. And everybody is almost certain to carry one bad gene. To reduce the incidence of bad genes by attempting to eliminate the gene from the population is therefore impractical even for the most common genetic diseases.

Little can be said about what costs would be involved in screening programmes for the diseases described earlier until reliable tests are available to detect the carriers and to detect in the foetus two copies of each deleterious gene. A convincing example of the benefits of screening outweighing the costs has been given earlier for the PKU screening programme in Britain. In an idealized cystic fibrosis screening programme, carriers would be identified and every marriage at risk would then be kept under observation. One such marriage would be identified for every 552 wives and husbands screened. (529 wives, 23 husbands; see Table 4.6.) Then there is the cost of at least one amniocentesis and the cost of an abortion. This has to be less than the cost of looking after a child suffering from cystic fibrosis for the duration of his or her life if it is to be economically worth while. If we assume that each screening test costs £5, probably an unrealistically high figure, and each amniocentesis £50, with the abortion possibly coming to £100, then the total cost of preventing the birth of one child would be about £3000. The cost of looking after a cystic fibrotic child would of course depend on how long he lived, but merely looking hard-headedly at the costs without including a factor for the human suffering, the balance would be in favour of abortion if the cost of looking after such a child, assuming an average lifetime of 15 years, came out to be more than £200 a year.

There are, however, other factors to be taken into account. So far it has been assumed that it is a matter of preventing the birth of an affected child by amniocentesis followed by abortion. But little attention has been given to research aimed at prolonging the lives of sufferers of these diseases and on making their lives easier. The life-expectancy of a cystic fibrotic child is increasing every year. So it is not possible to tell parents that their child will not see adulthood, for no one knows what the life-expectancy will be in 15 years' time. Developments in drugs might also make it easier for the child to get through the difficult early years, when he is particularly liable to suffer from chest infections.

In a similar way it must be shown that the costs of screening mothers-to-be for evidence of a PKU gene, followed by amniocentesis and possible abortion, must be less than the present costs of screening all newborn children for the disease. In this case the parents can be given the choice of abortion or of bringing up a child suffering from PKU. Although the success of treating affected children with a special diet is more or less proved, the parents might decide on abortion and another pregnancy to ensure that a normal child (although possibly a carrier of the disease) would be born.

Above all, perhaps, the one consideration which must be given greater attention is the possible effect of counselling and education on identified carriers of recessive diseases. Do screening programmes generate unnecessary anxiety among perfectly normal parents? Will parents faced with the prospect of producing affected children and carriers of the disease prefer in future not to have their own children but to adopt? Or will such parents prefer not to bring up children? Only time will tell what effect counselling will have on the numbers of affected children which are born every year.

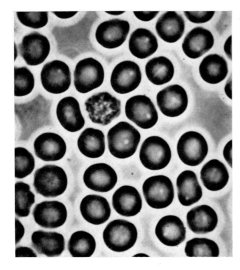

b, c. Red blood cells of an individual with the sickle-cell trait

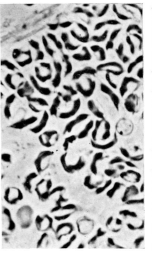

3b. Red cells in their normal disc-shaped state in the presence of oxygen

3c. The same cells when oxygen has been removed; they now have the characteristic sickle shape

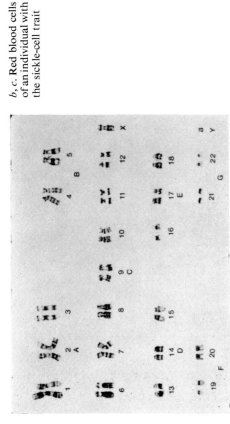

3a. The chromosome complement of a normal human male. The chromosomes (except for X and Y) are arranged in pairs and ordered by size following a standard convention. The numbers identify the 22 different chromosomes; the groups put together chromosomes of similar shape and size. The preparation shown had been treated using a technique that gives characteristic banding patterns. Each chromosome can be uniquely identified by its banding patterns. Pairs of homologous chromosomes (one from the mother and one from the father), other than X and Y, have the same pattern. Chromosome 21 is the one found in triplicate in Down's syndrome or mongolism. Rare diseases caused by three copies of chromosomes 13, 14, 15, or 18 and very occasionally 16, are also known but no other chromosome has ever been found in triplicate at birth. Loss of the Y chromosome, giving rise to XO (Turner's syndrome), is compatible with life but no other chromosome than the X has ever been found in a single copy at birth.

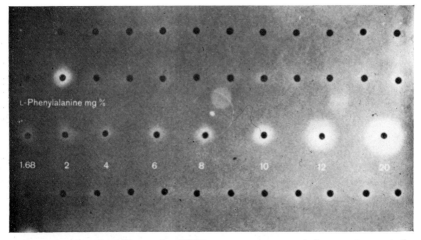

4*a*. Screening for phenylketonuria (PKU): an agar tray used for carrying out the Guthrie bacterial inhibition test. Each black spot is a disc impregnated with a blood sample to be screened. The third row is a control series of discs impregnated with various known concentrations of phenylalanine. One positive result can be seen in the second row (second from left). The positive reaction is indicated by a 'halo' representing inhibition of growth of the bacteria in the neighbourhood of the disc.

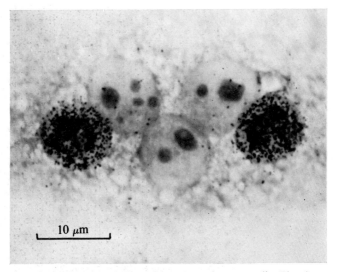

4*b*. A cell formed by fusion of human and mouse cells. The three nuclei in the middle are from the human cell parent and the two nuclei at either end, covered in dark spots, are from the mouse parent. The spots are due to the incorporation of a radioactive tracer into the mouse cells in order to distinguish them clearly from the human cells.

5 Organ transplantation

TISSUE grafting or transplanting is not a development of twentieth-century medicine. In fact, records are available of grafts carried out as long ago as 3500 B.C. by the Egyptians, but it was not until the sixteenth century that the first proper description of some of the techniques was made by Gasparo Tagliacozzi, a professor of anatomy at the University of Bologna. Not surprisingly perhaps, the theologians of Tagliacozzi's time attacked him bitterly, accusing him of impiously interfering with the handiwork of God and attributing his success to the intervention of the 'Evil One'.

Concern about grafting did not end with the work of Tagliacozzi, and it has persisted up to the present day. Corneal grafts received a great deal of publicity in the 1950s. Later, concern was expressed about kidney transplantation, and towards the end of the 1960s heart transplants made headline news. Between the sixteenth and twentieth centuries, tissue grafting was always an emotive issue; for example, towards the end of the eighteenth century the University of Paris banned all tissue-grafting operations.

In the absence of appropriate medical treatment any organ grafted arbitrarily from one person to another will survive for only a short time. Skin grafts, for example, will last only 10 to 15 days unless they are made between identical twins, when the graft will survive indefinitely. Genetic differences between people result in any transplanted tissue or organ being recognized as a foreign body by the recipient's immune system. Grafts are rejected for the simple reason that people are different from one another.

When the body is invaded by infection, the immune system

attempts to reject the infection, thus protecting the individual. But the immune system is a liability when foreign tissue grafts are involved. The differences between people which cause rejection of foreign tissue are rather similar to those which cause people to belong to different blood groups. No blood transfusions are undertaken unless the donor and the recipient's blood have been tested to see if they are compatible. The immune system is, however, much more complicated than the blood grouping system; it involves many more factors and its properties have still not been completely revealed.

Because of this incomplete understanding of the immune system, two approaches have been taken to maximize the chances of a graft taking successfully. The first is to try to inhibit the immune system by treating the patient with drugs. The second is to try to match as many of the parameters of the immune system of the donor and recipient as is possible before the graft takes place.

Even though the matching is not completely understood and drugs are still being developed, some tissue transplants have been successful enough to make them a reasonable form of therapy. In particular, corneal and kidney grafts have been successful, and the success of heart transplants continues to improve.

Tissue or organ transplantation may seem to be only distantly related to the other topics discussed in this book. But the legal, ethical, and moral problems associated with organ transplants are closely allied with those associated with these other topics, in particular, gene and cell therapy, which are discussed in Chapter 6. In these fields of research, biologists are attempting to replace single defective groups of cells in the body by other normal cells or to go to even finer detail and replace a single gene by another gene. Gene and cell transplants are both at a very early stage of research, but transplants of kidneys and grafts of corneas are already well-established and successful medical practices.

Heart transplants, which society was suddenly brought to face in 1967, had a deleterious effect on the progress of other transplants in Britain. It is an arguable point whether or not heart transplants were premature in the medical sense, but the supply of kidneys from cadavers decreased in Britain in the wake of the publicity accorded heart transplants. Thus kidney transplantation, a very successful surgical technique, suffered a setback and did not perhaps develop as fast as it would have done because of a reluctance by relatives to authorize surgeons to remove kidneys from cadavers.

During the 1960s, livers, lungs, and thymuses were also transplanted in humans, but up to 1973 these transplants had not been as successful as kidneys. There were encouraging signs at the end of 1973, however, that both liver and thymus transplants were becoming increasingly successful. More than 10 patients who had undergone liver transplants were still alive at that time, the longest-lived among them being a woman who underwent an operation in February 1969. Several successful thymus transplants had also been achieved by the end of 1973, and there was hope that this would also develop into an acceptable therapeutic practice.

The surgical techniques for transplantation can, in most cases, be developed with practice but the essential difficulty is to prevent the transplanted organ being rejected without leaving the body open to all sorts of infectious diseases. This is because the drugs used to counter rejection also suppress the body's immune response to infections. There is, however, some evidence that livers are much less susceptible to rejection than hearts and kidneys.

Kidney transplants and kidney disease

Kidney transplants have been very successful in recent years; more than 10 000 transplants had been carried out by the end of 1973. While the problems of rejection have by no means been completely solved, they are well under control. If the donor is a blood relative then there is at least a 75 per cent chance that the graft will still be functioning 2 years after the transplant. The chance of survival is especially good if the donor and the recipient are well matched. This is much easier to achieve for relatives than it is for randomly chosen pairs. There was a person living in 1973 who had received a kidney from a relative in 1958. On the other hand, if the kidney is taken from an unrelated dead donor, then there is about a 50 per cent chance that the kidney will be working in 2 years. Up to 1973, the longest functioning graft had been undertaken in 1964. With improvement in treatment, however, the chances of a graft from an unrelated dead donor functioning at the end of 2 years are increasing.

What is the prognosis for patients who do not acquire transplanted kidneys? Fatal kidney disease is a relatively common affliction of young people. In Britain, between 1500 and 2000 people between the ages of 5 and 55 die every year from kidney malfunction. Many of those who die are teenagers.

There are two ways of treating patients suffering from kidney diseases. The patient can be put on a kidney machine or he can be given a new kidney from a relative or an unrelated donor, usually a cadaver. A kidney machine mimics the action of a real kidney, filtering the blood to remove waste products. The machine achieves this by taking the blood from the body and filtering it through artificial membranes. The purified blood is then returned to the body. The regular use of a kidney machine is referred to as recurrent maintenance dialysis.

In treatment by dialysis with an artificial kidney machine the patient must be connected to the machine for two or three sessions of 12 to 14 hours each week. Since the blood circulates through the machine, connections must be made between the blood-vessels and the artificial kidney each time. Complications can occur at the connecting sites, and periodically these sites have to be changed because of infection. Between treatments, the patient has to restrict the amount of fluid he drinks and the salt he eats. The person with defective kidneys who opts for dialysis must attach himself to the machine regularly for the rest of his life or until he is given a transplant. Connections with the machine must be made with the help of relatives or friends if the patient is at home, or with the medical and nursing staff of a hospital. Life for the patient is considerably limited in both the physical and psychological senses, but he or she can lead a relatively normal life between treatments. Certainly, treatment with an artificial kidney is immeasurably better than no treatment at all.

The long-term prognosis for patients on dialysis is good, provided there is adequate supervision, and complications are rare. But with up to 2000 cases requiring treatment every year it would not take long for all these people to become a huge drain on both medical and nursing resources. In Britain, the financial burden on the National Health Service would in a few years become too great. The costs of dialysis in the United States in 1972 were estimated to be between $10000 and $25000 a year for each patient. In Britain in 1973 the figures were £2000 per patient a year for home dialysis and £2500 for hospital dialysis. This does not take into account the £3000 needed to set up a dialysis unit at home.

It would therefore seem prudent to try and increase the number of patients being maintained at home to decrease the cost and also to release hospital staff for other duties. Maintenance dialysis in Britain moved from being a research tool to a practical treatment in the mid-

1960s, when advice was given to the Department of Health and Social Security by a committee of physicians and surgeons which met under the chairmanship of the late Lord Rosenheim. In 1965 there were only a few units in Britain carrying out maintenance dialysis and these were chiefly orientated towards research. By 1970, there were 31 major units in England and Wales, 6 in Scotland, and 1 in Northern Ireland. There was also a unit in London which was outside the National Health Service and provided private training facilities for home dialysis.

In 1967 the increasing cost of dialysis and its success as a technique made the Department of Health and Social Security advise that home dialysis be made more widely available. So, by June 1971, 24 out of 39 major units were undertaking such a service. Since then a further 6 units have offered a home service.

The success of the measures to make dialysis more readily available and to make it available in the home is seen from the fact that fewer than 200 patients in England and Wales were on maintenance dialysis in 1967, but in June 1973 the figure was 1509. In 1967 most of the 200 were treated in hospital; in 1973, 964 were being treated at home and 545 in hospitals. In Scotland, with 4 kidney units, 44 patients were accepted for maintenance dialysis in 1968-9 and 68 and 40 respectively in subsequent years. In Northern Ireland there were 4 patients on maintenance dialysis in 1967 and 19 in 1971. All these were maintained in hospital.

But the number of patients on dialysis is not nearly enough to reduce the number of people (almost 2000) who die each year from kidney diseases in Britain. Between 1 July 1972 and 30 June 1973, 630 new patients started dialysis treatment; during that year the number on home dialysis increased by 192 and the number in hospital on dialysis went up by 23. The number of kidney transplants, on the other hand, is not growing at a fast enough rate. In 1969, 200 transplants were carried out in kidney units in the British Isles. In 1970, the figure was 274; it was 315 in 1971 and 465 in the year ended 1 February 1973.

Kidney transplants have been more successful than other transplants—with the exception of corneal transplants—chiefly because of the existence of dialysis as a back-up method of treatment should the transplant fail. Dialysis is also essential before a kidney is transplanted, in order to keep the potential recipient fit until a suitable kidney becomes available. It turns out that even if a first kidney transplant fails, the probability of success with a second transplant is not necessarily reduced. This in itself is rather surprising, for with normal immunological responses such a second or subsequent graft would be much worse off than the first. But the use of immunosuppressive therapy appears to prevent this usual secondary response.

Kidney transplants became an accepted form of treatment for chronic kidney failure in Britain in 1967, two years after dialysis was accepted as treatment. In July 1972 there were 14 units in Britain where kidney transplants could be carried out, and this number is to be increased to 20 by the mid-1970s. At the end of 1973, however, none of these centres was working near capacity because they could not obtain enough kidneys for transplantation. For example, the kidney unit at Addenbrooke's Hospital at Cambridge carried out 50 kidney transplants in 1971, but the unit could comfortably have coped with at least 60, so far as the supply of beds, nurses, and dialysis was concerned.

The cost of a kidney transplant in Britain was estimated to be about £1500–£2000 in 1973, whereas in the United States it was estimated to be between $10000 and $25000 (the same as the cost of dialysis), but if there are complications following the operation the cost could escalate. The cost of the transplant operation is, however, decreasing each year as more and more surgeons gain expertise in the work, according to the U.S. Department of Health, Education and Welfare. Economically it is clear that a successful transplant is much better than a long period of dialysis.

How do the chances of a patient surviving depend on the treatment he is receiving? And how does his quality of life change with treatment?

In July 1972 a joint committee under the chairmanship of the late Lord Rosenheim, which had representatives from the Royal Colleges of Physicians, Surgeons, Obstetricians and Gynaecologists, and Pathologists, the Royal College of Physicians of Edinburgh, the Royal College of Physicians and Surgeons of Glasgow, the British Paediatric Association, and the Renal Association, surveyed the performance of 29 of the 42 dialysis units then operating in Britain. Their report shows that 64 per cent of patients undergoing dialysis survived for 3 years (the extent of the survey). Of the patients that underwent a combination of transplantation and dialysis—with dialysis before and after the transplant if the transplant failed—59 per cent survived for 3 years. The difference in survival of these two groups is therefore not significant.

The chance of a kidney graft being successful for 3 years, on the other hand, is 41 per cent according to the survey. This is not to be taken as the chances of survival of the patient, for if the transplant is unsuccessful then the patient is returned to dialysis and may be given another transplant.

The probability of survival for 3 years of a patient undergoing either dialysis or transplantation is therefore about 60 per cent. The chance of a patient with chronic kidney failure surviving for the same time without treatment is, of course, usually nil. For patients on dialysis machines or with patients who have had transplanted kidneys, the chances of living longer than 3 years cannot be assessed because large-scale treatment in Britain, at least, has been available for only a few years. Some statistics are illustrated in Fig. 5.1.

The statistics, however, hide the important fact that the two groups—those undergoing maintenance dialysis alone and those having had a kidney transplanted as well as receiving dialysis treat-

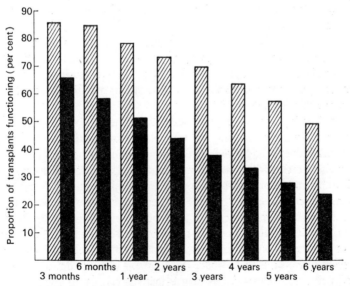

FIG. 5.1. Average survival time of kidney transplants carried out between 1967 and 1971. Shaded columns, transplants from sibling donors (1170 cases); solid black columns, transplants from cadavers (4348 cases). (Adapted from *Eleventh Report of the Human Renal Transplant Registry*, 1973.)

ment—are not strictly comparable. The 41 per cent chance of a kidney graft being successful after 3 years underlines the fact that without facilities for a return to a dialysis unit the survival of patients who had undergone kidney transplants would be lower. But there is little doubt that the quality of life of a person who has a transplanted kidney is immeasurably better than that of a person who has to undergo regular maintenance dialysis.

The encouraging aspect of kidney transplants is that the survival rate continues to increase, as it has done during the past few years. But the survival rate for patients on dialysis is constant with no increasing trend.

The statistics also reveal another encouraging aspect of transplantation. This is that the chances of survival are much greater if the patient is under 35. Somewhere between the ages of 35 and 45 the advantages of transplantation appear to be outweighed by the advantages of dialysis. The survey of the British dialysis and transplant units shows that for some men under 35 the chances of being alive 3 years after the transplant are as high as 70 per cent.

Of the 1500 to 2000 patients who need treatment for chronic kidney disease every year in Britain, 1095 received treatment between 1 July 1972 and 30 June 1973. There is still therefore a large gap between the demand for kidney treatment and the availability of such treatment, although it narrowed a great deal in the late 1960s. The gap is larger than appears from these figures, however, for the statistics do not necessarily refer to patients whose diseases have become evident after mid-1972.

As already stated, one of the reasons for the gap between demand and supply of treatment by transplantation is the shortage of cadaver kidneys. Kidneys for transplantation must be removed from the dead person within an hour of death and then can be stored for 12 hours on ice, or for 24 hours on a perfusion machine before being transplanted. There is clearly a need to develop methods of preserving kidneys for even longer before they have to be transplanted. Not all the kidneys that are obtained can be used. To ensure that the transplant has the maximum chance of success, a series of tests is carried out to compare the recipient with the donor.

As mentioned earlier, there are factors, similar to blood-group differences, which are not the same in donors and recipients, which are responsible for rejection. Matching these factors should increase the chances of a graft surviving. It turns out that the system of blood-

Organ transplantation

group matching (the ABO system) used for blood transfusion is also important for transplantation of kidneys. The first step before proceeding with a kidney transplant is therefore to ensure that the blood groups of the donor and recipient are compatible.

In the 1960s, a complex histocompatibility system (the HL-A system as it is called) was discovered which has turned out to be the most important apart from the ABO for the survival of transplants. The matching for this system is done with the white cells of the blood rather than with the red, as is done with the ABO system. The HL-A system is complicated, and it is almost impossible to find a donor and a recipient with identical characteristics. The best match possible must therefore be used, even if it is not ideal. Clearly the chances of obtaining a match will be greatest if there is a large pool of potential recipients for comparison with the donor.

A disadvantage of treating the recipient with drugs designed to suppress the body's natural inclination to reject the grafted tissue is that the body's resistance to all kinds of infection is lowered. It would be a great step forward if a method could be developed that would suppress the body's tendency to reject the transplanted organ or tissue but would leave resistance to infection unchanged. One approach of research towards this end is based on the injection into the recipient of factors obtained from the blood and other tissues of the donor. The aim is specifically to inhibit the recipient's natural response by the presence of the appropriate factors from the donor. By the end of 1973 there was no reported success of this kind of work.

Why are so few kidneys available to the surgeon? The chief reason seems to be apathy on the part of society to the need for kidneys after death. In 1972, 7800 people were killed in Britain in road accidents: a source potential of 15600 kidneys. Many other people die of brain tumours and brain haemorrhages, all of them potential kidney donors. But of this enormous potential pool of kidneys, only a few hundred become available for transplantation.

In part, the shortage of kidneys is a consequence of the legal situation in Britain, where permission to remove organs must be obtained either from the donor before his or her death, or from the relatives. It is argued that the medical profession itself is also to blame for the shortage of kidneys. After tending a patient who dies, the doctor has no wish, understandably, to become involved in difficult interviews with usually distraught relatives to ask their permission to remove kidneys from a body. The doctor's work does not end there, for he then has to contact the nearest transplantation centre and advise it of the availability of a suitable kidney.

In Britain there is a national centre at Bristol, which co-ordinates kidney transplants. A computer there stores information about patients all over Britain who are in need of kidneys. Also stored with the names are their blood types and histocompatibility classification. The recipients are also classified according to how urgent their need for a transplant is.

The local transplantation centre which has been notified by the doctor is responsible for the 'typing' of the donor, that is, analysing the blood group and identifying the factors mentioned above. These are fed to the computer, which compares the data with those in its store and then lists the names of potential recipients. All further details such as arranging for the transport of the kidneys and informing the doctors in charge of the recipient, are the responsibility of the National Centre.

As mentioned previously, kidney transplantation in Britain has probably suffered in recent years because of the adverse publicity given to heart transplants. The public, after the massive public relations exercise which accompanied the first few heart transplants in South Africa, the United States and Britain, appears to have become less enthusiastic about all forms of transplants, including kidneys. The problem is basically twofold. First, it seems that relatives shy away from giving permission for organs to be removed from their dead next-of-kin because they fear publicity. Second, there is some disquiet in case organs should be removed from donors prematurely, that is, before death has occurred.

But how is death defined? Artificial aids to breathing and circulation complicate matters, but in Britain death is certified if the spontaneous circulation of the blood and breathing have come to an end and cannot be restarted. In most European countries and in the United States, on the other hand, death is held to have taken place when brain activity stops. The confidence of some people in doctors is low and must be restored so that there is no fear lest death is certified prematurely. In this way, confidence in kidney transplants will be rightly built up. There is no doubt that this surgical technique can restore a normal life for many years to many sufferers of potentially fatal kidney diseases.

Corneal transplants

Corneal grafting or transplantation is a surgical technique which is chiefly used to cure blindness caused by scarring of the cornea. The transplants, however, are not carried out solely for this purpose and in Britain about a quarter are done as a therapeutic measure, to relieve intractable ulcers for example. This is an affliction which primarily affects the young rather than the old. There are between fifteen and twenty million blind people in the world, and in England and Wales in 1972 103 853 people were registered as being blind. The number of blind people is almost constant, about 12 000 new blind people being registered every year and a similar number either dying or having their sight restored. Corneal grafting, it is estimated, would give sight to between 1500 and 2000 people a year in Britain.

Corneal transplants have a longer successful history than any other kind of transplant. The first successful corneal graft between humans was accomplished in 1905 by Dr. E. Zirm in Olmutz, a small town in Moravia. Zirm grafted the cornea of a boy, whose eye had been removed in an accident, to a man whose eyes had been affected by lime. Several attempts at grafting had previously been carried out with corneas obtained from animals, but little success had been obtained. Between this time and the start of the Second World War very few corneal grafts were carried out, chiefly because success could not be guaranteed, but in 1937 Dr. V. P. Filatov of the Soviet Union used eyes from cadavers as a source of replacement corneas and also developed techniques for corneal transplants which were superior to those in use before then.

The Second World War, with its many eye injuries, combined with the work of Filatov, resulted in corneal grafting coming into its own, and now more than 1200 corneal grafts are carried out every year in Britain. In the 1930s barely half a dozen grafts a year were carried out in London hospitals, compared with almost 400 a year in the early 1970s at one London eye hospital. But in spite of the massive increase in the number of corneas transplanted every year, only about half of the required number are being carried out.

Compared with kidney transplants corneal transplants are much more successful. Over 80 per cent of corneal grafts 'take' and rejection, if it is to occur, will generally take place between the first and fourteenth day after the operation. There is some evidence, however, that rejection occasionally takes place up to 2 years after the operation. No complications are expected to occur in most cases after 3 weeks. Rather as with kidney transplants, if the first graft does not take, then a second or subsequent graft has the same chance of success as the first one and rejection need not follow. But, of course, the patient in this case is in no immediate danger of dying and the eye itself is seldom lost.

Corneal grafts differ in yet another aspect from kidney and other grafts in that general treatment of the patient with immunosuppressive drugs to prevent the body rejecting the cornea is not usually considered necessary. But when rejection takes place after 2 weeks, then such drugs are successful in preventing it proceeding further. The reason for this is that there are few blood-vessels in the cornea and the body therefore does not reject the tissue as violently as it rejects foreign tissue grafted to other parts of the body. To prevent rejection occurring after 2 weeks a more detailed matching of tissues, similar to that for kidneys, would be advisable to minimize the effects. Up to the end of 1973, no such large-scale matching of tissues had been considered necessary for corneas.

The reason why fewer corneal transplants are carried out than are needed is that there is a shortage of qualified eye surgeons as well as a shortage of eyes. The availability of eyes, like kidneys, is governed by the Human Tissue Act of 1961, but eyes are easier to obtain than kidneys. Eyes can be removed from the cadaver up to 12 hours after death, compared with the much shorter time-limit for kidney removal. Eyes remain viable after being stored for weeks in a refrigerator (compared with the 24 hours that the kidney can be stored on a perfusion machine), although for the best results it is advisable to use the eyes within a few days of the donor's death.

Although eyes may be considered by the surgeons to be difficult to get hold of in Britain and other developed countries, the difficulties are small compared with those faced by surgeons in some developing countries. In some of these countries, even though there are no legal barriers, religious, cultural, and social barriers to the removal of eyes from cadavers are almost insuperable. In some eastern countries, such as Malaysia, it is not officially permitted even to be the recipient of a graft. The usual solution in most of these countries is to fall back on eyes delivered from an eye bank in some other country, where the public are more prepared to give their eyes for grafts.

An eye bank is no more than a collection, storage, and delivery service of eyes which has at its disposal a suitable refrigerator. The first eye bank was opened in the United States in the early 1940s and now there are no less than 80 such banks there. Another eye bank was started at the Westminster Hospital in London in 1961, which became the Westminster–Moorfields Bank in 1964 and joined with an existing bank at the East Grinstead Hospital in 1967 to become the British National Eye Bank. There are officially 6 other regional banks in Britain. About one in five of the eyes obtained by the British National Eye Bank is sent abroad and more than twenty different countries have received eyes from Britain.

In principle, eyes could be stored indefinitely at liquid nitrogen temperatures ($-196\ °C$) thus making better use of available eyes, though at some additional cost. The danger, in theory at least, is that the tissues would be damaged during the freezing or thawing processes. But in Britain at least, eyes, while not plentiful, can be obtained and it has been thought unnecessary to store them for much longer than a few weeks. If it turns out that matching of tissues to prevent long-term rejection is sometimes necessary, then it would clearly be advantageous to have a large store of frozen eyes so that as good a match as possible could be made between the tissues of the donor and those of the recipient.

The success of corneal grafts from cadaver eyes also makes it less necessary to develop artificial corneas. They have been manufactured and tried in eye surgery but they are difficult to attach to the eye and there is a tendency for them to pop out under pressure of the eye fluid.

Transplanting whole eyes is simply not practicable at present. It would be difficult, if not impossible, to maintain a continuous blood supply to the retina during the operation, and the technical problems of such an operation also appear insurmountable. Surprisingly, the difficulty of connecting the optic nerve is not considered to be a critical problem: the brain would be expected to interpret what the eye sees.

Legal aspects

During the passage through Parliament of the Corneal Grafting Act in 1952 (an Act which was supplanted by the Human Tissue Act in 1961), an attempt was made to increase the supply of eyes for transplanting by persuading people to donate their eyes for medical or research purposes after death. As a result of this campaign, which was carried out through newspapers and radio, there are 60000 pledges in filing cabinets in a basement in London. These, it seems, are merely gathering dust, for they have hardly ever been consulted by eye surgeons. This exercise can be compared with the campaign launched in December 1972 by the Department of Health and Social Security to persuade people to 'contract-in' to a kidney donation scheme. The scheme is not likely to produce a flood of volunteers, but it is hoped that it will provide the necessary publicity for kidney transplants, just as the eye donation scheme did for eye grafts in the 1950s in Britain. The message the surgeons wish to put over is simply that people dying from kidney diseases or suffering from blindness deserve to be given priority over corpses.

In Britain the transplantation of organs is controlled by the Human Tissue Act of 1961. In spite of the efforts of a group set up under the chairmanship of Sir Hector MacLennan in 1969 to advise Parliament on possible ways of amending the Act, the law is still the same today as it was then. This is chiefly because the advisory group was evenly divided in its recommendations between contracting-in and contracting-out and no clear route was left open for the then Minister of Health to follow.

The 1961 Act provides that organs can be donated for therapeutic purposes or research by a decision in writing at any time, or by an oral declaration in the presence of two witnesses during a last illness. The person lawfully in possession of the body after death may execute such decisions by authorizing the removal from the body of any part so donated. The Act also provides, without prejudice to the case where the deceased has made an explicit decision to donate organs, that the person lawfully in possession may authorize the removal of parts. This he can do if, having made such reasonable inquiry as may be practicable, he has no reason to believe there was objection on the part of the deceased before death or that the surviving spouse or any surviving relative would object.

Assuming this permission can cause difficulties for the surgeon. During the summer of 1973, a teenage boy was killed in a road accident in Britain while his parents were on holiday abroad. The surgeon removed the boy's kidneys without parental permission, but the resultant publicity and outcry following the parents' reaction when they found out about the removal of the kidneys had yet another detrimental effect on the supply of kidneys for transplantation.

It would seem that there is a case for stating the law in less ambiguous terms, even if it is felt that no action should be taken to change it radically in order to increase the supply of kidneys for transplantation. But there is no clear mandate for this in the medical community, and some members of the MacLennan group felt that if a radical change of the law was not acceptable to Parliament, then a wider interpretation of the current law would be better. The law can be interpreted in the interests of transplantation if, for example, 'the person legally in possession' is taken to be the hospital or its officers at the time of death and if 'reasonable inquiry as may be practicable' is taken as a flexible commonsense statement that doctors should try to obtain the consent of relatives within the limited time available for the removal of the organ. The cause of transplantation will also benefit if 'surviving relatives' is taken to mean only close relatives. If these phrases can be so interpreted, then the law would be tolerable from the point of view of surgeons and many more kidneys would be available for transplantation. But, as pointed out by the MacLennan group, such an approach would have its advantages and disadvantages. The merit of such an approach would be the flexibility inherent in the law, but the disadvantage would be the lack of clarity for both the public and the professions. There would also be some risk of legal action against a doctor who had acted in good faith.

Although the MacLennan group did not present a unanimous view on the way in which the law should be changed, it did present a series of general conclusions which the group considered to be of cardinal importance.

The group agreed that a reliable and proved technique of saving life, such as kidney transplantation, should not be held back by obsolete laws or by lack of information which might affect the attitude of the medical and nursing staffs of transplant teams.

It also agreed that there is a need to increase the supply of organs for transplantation in order to enable surgeons to treat more patients and to improve the prospects of survival, simply because a greater supply of kidneys increases the probability of a good match being maintained between the donor and the recipient.

The group also agreed that there must be safeguards for kidney transplants, but, of greater importance, the public must be made aware that there were safeguards. In part, these safeguards manifest themselves in the traditional ethical standards of the medical and nursing professions.

In particular, the group also made a recommendation to ensure that there would be no accusations that organs were removed from donors prematurely. The doctor responsible for the donor is always different from the transplant surgeon, but according to the report two doctors should certify death, one of whom should have been qualified for at least five years; both should be independent of the transplant team, and the certification of death should be made without regard to the possibility of transplantation. When a patient was being maintained by resuscitation apparatus, the decision whether to carry on with the artificial aid or to discontinue such support should also be made without regard to the possibility of a transplant. The group recommended that the two doctors who certify that the patient is dead should record their observations independently on a specially provided form, which would be kept with the case records and would be available to the coroner, or, in Scotland, the procurator fiscal. This specific proposal has not been implemented.

Another safeguard which the group emphasized is that each person must have a right to make a binding decision as to whether or not he wants his organs used for transplantation after his death.

Any changes to the law must be so framed, said the group, that they can be applied to the transplantation of all organs and not only kidneys. The way in which a person wants to dispose of his own organs after death must have overriding consideration, which generally is the position under the 1961 Act.

The MacLennan group was also worried about the publicity accorded to transplant operations. Publicity can distress recipient patients and relatives of donors. It must be realized, said the group, that patients immediately after the operation have still a long way to go to full recovery and so they should be spared needless distress caused by publicity. The group commended the practice of hospital staffs in refusing to give the names of either recipients or donors to the press and it also praised newspapers who respected the privacy of the people concerned.

The group's final general conclusion was that the co-operation of coroners and procurators fiscal was essential if a supply of organs was to be maintained. The group pointed out that, where appropriate,

Organ transplantation

coroners could rely on authoritative pathological reports from hospitals so that the body could be disposed of before any inquest.

These points raised by the MacLennan group are still of great importance, but subsequent events have emphasized some of the issues, in particular those of keeping the names of patients and donors secret. Not only in transplantation cases does the need for secrecy arise, but also in most of the other emotive topics discussed elsewhere in this book.

But how can a greater supply of kidneys be obtained without infringing upon the right of a person to decide for himself whether or not he wants his body put to use after his death? There are several possible answers to the question, which the MacLennan group classes as follows.

(*a*) There should be no change in the law but, as mentioned earlier in this chapter, the law should be reinterpreted in order to benefit transplantation.

(*b*) The law should be amended slightly to remove the ambiguities inherent in the 1961 Act. In this way the 'person lawfully in possession' of the body would be defined as the hospital authority during the time between death and the time when the next of kin or executors claimed the body. The term 'surviving spouse or any surviving relative' could be replaced by 'next of kin' and a definite time be allocated to carry out all 'reasonable enquiry as may be practicable'.

(*c*) A single public register should be set up which would record both objections and consents to an organ being removed from a person in the event of his or her death. Apart from providing a pool of potential donors, the MacLennan group saw such a register as an indicator of public opinion in the country that would enable a subsequent decision to be made on whether to continue with such a double register or merely to keep records of potential donors or objectors alone. The disadvantage of such a register is, as pointed out by the group, that the position of those who have not registered either an objection or consent would be unclear.

(*d*) A single register should be kept of people who wish to 'contract out' of donating their organs after death. The surgeons would then be free to remove any organ from a corpse provided that there was no record in a central registry that the dead person objected. Such a system as this already exists in Sweden, Israel, Italy, and France.

The Human Tissue Act of 1961 is based on a 'contracting-in' principle: no organs can be taken unless permission is already at hand either from the deceased person or persons in charge of his body. In 1969 an abortive attempt was made in the British Parliament to introduce a private member's bill which, as well as introducing some of the general recommendations of the MacLennan group mentioned earlier in this chapter, would have changed the system in Britain to a contracting-out system. The failure of the bill to pass through was an indication that Parliament, at least without a strong government lead or guidance, was not then ready to accept such a scheme. Six of the eleven members of the MacLennan group, however, favoured a contracting-out scheme while the other five recommended a limited amendment to the law as in (*b*) above. But all the members of the group said that they would favour introducing a double contract, as in (*c*) as an experiment, together with widespread education and provision for cohort enrolment.

The Department of Health and Social Security in December 1972 put its weight firmly behind the present law by introducing a contracting-in scheme designed not only to increase directly the number of prospective donors, but to give much-needed publicity to the need for kidney donors. The present policy of the D.H.S.S. is that there should be a contracting-in either by the patient or by his relatives in the event of his death.

But in the present scheme the D.H.S.S. has not set up a central registry of potential donors. The rationale behind this, it seems, is that the hospital authorities can rarely be absolutely certain of the identity of an accident victim in the short time between death and the time when an organ has to be removed. The chief purpose of this scheme is that it will provide the necessary publicity for the need for kidneys. It will also produce a background against which further campaigns might be based. It could, however, be used as an excuse to do nothing further and the very limitations of the scheme could reinforce public apathy rather than strengthen concern.

In this way, the D.H.S.S. has taken up in a small way the advice of the MacLennan group on the need for publicity to increase the supply of kidneys for donation after death. The group's recommendation that prospective donors should carry a card which would be 'durable, immediately recognizable and unique' has also been taken up.

Ethical aspects

The ethical and moral issues surrounding transplantation have created a great deal of discussion in recent years. The major Churches have had different thoughts on the subject, although at present there is no dogmatic opposition to transplantation within the major Christian theological and ethical tradition.

Some of the ethics of medical science and practice in the western world have been formulated in terms of positive law; for example, the English common law and statute law and the Code Napoléon; others remain as rules or principles of practice held corporately by the profession with sanctions of various weight behind them.

This ethical tradition attributes finality and totality to each person and so society is reluctant to use people without their consent as a means to an end, or to subordinate them to any 'totalitarian' claim. The tradition also allows people to fulfil themselves in social or community relations, but while a high degree of individualism is allowed, the Hippocratic–Judaeo–Christian tradition does not allow any one person an absolute claim when he seeks to isolate and assert personal interest over and above the common interest.

Ethics develop out of the tension between the legitimate claims of persons and those of society; that is, between the individual interest and the common good.

Codes of practice for the medical profession regard the patient as the primary object of medical care and these codes seek to assure that no interest of the patient is subordinated to other interests within some degree of safeguard and consent. But in practice no absolute protection is afforded to the patient on all occasions when this interest conflicts with the public interest. For example, the patient's interest is sometimes not given priority in a court of law or in medical certificates, when details of a patient's illness have to be revealed.

As a working rule, however, the patient's interests are assumed to come first and our ethical tradition is such that society assumes it has a duty to protect this interest. If there is a need to disturb this tradition and put some other interest before that of the patient, then it is up to the person alleging this to present the proof that this is necessary. This is the basis of medical ethics within western society and so it is the basis of the confidence and trust which the patient places in the medical profession.

The medical ethical tradition of the western world also assumes that man, although a product of nature, transcends it in a unique way and has not only a liberty, but a duty, to extend control over it. This duty includes a responsibility for the development of the human species in its genetic as well as in its other characteristics. Biblical ethics speak of man's covenanted obligations, implying that man may rule and control only when he himself obeys the laws to which he is subject. In his control over nature, therefore, man remains subject to it.

The giving or donating of human tissue for transplantation needs no elaborate ethical justification in that it is merely an apt expression of human mutuality or corporateness. In New Testament terms it merely exemplifies the expression 'members one of another' which developed from the Old Testament idiom 'bone of your bones and flesh of your flesh'.

The Roman Catholic Church has had difficulty over the ethics of the mutilation of the human body involved in the donation of tissue, which is not directed towards the good of the donor but towards another person. This, to some Roman Catholic moral theologians, appeared to breach the safeguards of bodily integrity but these difficulties have been shown to be surmountable in recent years.

Some Protestant churches, which readily accept acts done for the good of other persons, accept the giving of tissue more readily. They are therefore less sensitive to the need for safeguards which Roman Catholic moralists have sought to uphold.

It is, of course, necessary that a donor gives his full, free, and informed consent to the removal of any tissue from his body. This follows simply from the fact that the basis of moral community is the freedom to will and perform moral acts; and the word donor, of course, loses its meaning if consent is not obtained.

Superficially, however, the problem of removing organs from cadavers is simpler, although it is socially more complicated than removing organs from live donors. *Ex hypothesi*, a dead man has no interest in his own body, and as there is in law no property in a corpse the notion of 'consent' becomes detached and of apparently diminished exigency. Also, if in our ethical tradition absolute value were attributed to human life an overriding duty would be asserted to take any cadaver organ which might save, or prolong, a human life. Within this context no lack of consent or family opposition would be allowed to stand in the way.

In practice, however, only a very high value is attached to human life—not an absolute value. Thus public policy to make available cadaver organs for transplantation must, at present, be framed with this high value and not an absolute value in mind. The policy must weigh the interest of the patient in surviving against other proper interests of the community. And these interests include the minimizing of distress to bereaved relatives. In the long term, the public must be persuaded to do these acts—such as donating organs—voluntarily, and decisions made against the wish of the people involved must be minimized. These arguments are particularly relevant in considering whether a contracting-in or contracting-out scheme should be used to increase the supply of cadaver kidneys for transplantation.

A much more emotive topic is whether human tissues should be obtained by purchase or by a gift. This question is related to both consent and the voluntary principle. It appears to have practical consequences in the quality of blood obtained when donations are paid for, as happens in the United States. The possibility of the donor gaining either money or perhaps a remission of a jail sentence might lead him to hide facts of his medical history. Some would argue, nevertheless, that payment should enter a transaction for the surrender of an organ or a tissue. There are others who would argue that such a transaction should be considered differently. It appears that public opinion in Britain, at least, sees such a surrender as a gift rather than a transaction. In this respect the motives of donors where money is involved are questioned. It is extremely unlikely that in Britain either kidneys or corneas will be obtained for money. Live kidney transplants are invariably between relatives, while the supply of eyes also precludes the necessity of payment. It is, of course, possible that the relatives of dead donors might want payment, but this is another issue. As an example of the feeling about payment, in the spring of 1970 it was disclosed that an embryologist was buying foetal tissue from an abortion clinic for experimental purposes. There was an outcry about the sale aspect and the public was apparently blinded by this to the genuine research interest behind the purchase.

This raises the fundamental point that when the ethics of a particular act are being formulated consideration should be given to public attitudes, even if the act is innocent in itself.

It is difficult to have one set of ethical judgements covering the entire range of tissue transplantation, because some forms of transplantation, such as kidney transplants and blood transfusions, are accepted practices, whereas others are in the development and experimental stages—heart transplants being prime examples, although there was some evidence at the end of 1973 that such transplants were becoming increasingly successful.

When discussing transplants which are in the development stage, the ethics of the situation must be based on the concept of an acceptable risk to the patient. Seldom can the probabilities of success of any operation, let alone a transplant operation, be determined without some degree of trial and, by definition, without some degree of risk. This is very difficult ground and the chief problem is how to designate responsibility.

How is the decision arrived at as to which patient suffering from kidney failure is to be given a transplanted kidney? In Britain with its National Health Service, the question is asked with particular interest to see whether recipients come from all social classes or whether one social class is favoured over any other. Basically as far as the patient is concerned the decision as to whether or not he is given a kidney graft should be a medical one, based on the prognosis for the patient with and without a graft. In practice it also depends on how far away from a kidney transplantation centre a recipient lives. It is agreed, however, that the medical evidence should not be the only basis on which decisions are taken. The degree of surgical intervention should be based not simply on the patient's pathological condition (for example, his kidney failure) but also on the patient as a person living in a social context of his own. The patient's relatives, who may be very much affected by his medical condition, have also to be taken into consideration. But who makes these ethical decisions? The patient's medical practitioner clearly decides whether the patient's pathological condition will be better or worse after a kidney transplant, but is the practitioner also to weigh the other factors before deciding to proceed with the transplant? It is more than possible that a procedure which would be strongly recommended for one patient would not be suitable for another patient because of these extra-medical factors. The question to be asked is whether such extra-medical factors are related to the social class of the patient.

Yet another possible issue which might arise with kidney transplants derives from the shortage of suitable donors. A person with the necessary money could approach a surgeon on a private basis, in order to receive preferential treatment. The difficulty so far as the

Organ transplantation

surgeon is concerned is whether to accept the patient knowing that the kidney which he would receive could have gone to another sufferer, perhaps a more deserving one.

Another factor which must be considered when discussing transplantations is whether society can justify the operation taking place. The issue, in fact, is one of cost *versus* benefit. In some health and medical matters (see Chapter 4) it is obvious that the benefits of a particular form of treatment or preventative treatment outweigh the costs. The way to proceed is then clear. The limitation on kidney transplants in Britain at present is not the cost, but the shortage of kidneys. If this were not so then other factors would have to be considered, some of which do not clearly fit into a cost–benefit equation. One of these is the quality of life of a person who has undergone a kidney transplant, which is immeasurably better than that of a person on maintenance dialysis. Another factor, more easily quantifiable, is the productivity of a person after a transplant compared with what it was before. Unfortunately decisions on whether or not to proceed with a line of treatment cannot avoid these unpalatable financial issues. The question with kidney transplants is, in essence, whether a great deal of money should be spent to save a few lives, or whether it would best be spent elsewhere. As an example, more lives could possibly be saved by the outlay of the same amount of money in equipping accident units. The money could also be spent in improving the facilities at mental hospitals so that the mentally sick could lead a better life.

6 *Genetic engineering and cloning*

MORE than a thousand diseases caused by malfunction of the chromosomes or of the genes are known, but up to the end of 1973 only about a hundred of these could be detected while the foetus was still at an early stage of development in the womb. But not even all the diseases which can be determined before birth can be diagnosed early enough in pregnancy for the foetus to be aborted. How can the other vast number of genetically determined diseases be controlled? So far only a few of these diseases can be treated successfully by drugs, diet, or transfusion; and for most genetic diseases little can be done for the sufferers at present.

Another way of dealing with genetic diseases is to try and cure them at the most basic level by modifying the gene itself. If, for example, the defective gene which causes a certain disease can be changed into a normal functioning gene in either the sperm or the ovum, then an hereditary cure would be effected in the sense that the person would pass on normal genes to his or her offspring.

The procedure of replacing a defective gene in the sperm or ovum, however, produces no relief for the individuals suffering from the disease. In order to effect a cure in the sufferers themselves the gene must be changed in those cells or tissues which are not functioning properly and are therefore responsible for causing the disease. Consider, for example, a disease which expresses itself as a malfunction of the cells of the liver. If liver cells containing normal genes could be put into the patient's liver, the disease might be cured. This is a genetic cure, or 'gene therapy', in which the cells causing the disease have been supplanted by cells containing 'normal' genes. The genes in the patient's sperm would still be abnormal, for there is no way in which the normal genes from the inserted liver cells could transfer to the ovum or sperm cells.

The science of improving the hereditary qualities of future generations of mankind is called *eugenics*, and within this term is encompassed all methods of achieving these ends. An important aspect of eugenics is to prevent the birth of children with genetic disabilities (as discussed in Chapter 4). Responsible and informed parents will, for example, plan their families with such factors as physical and mental health in mind. Selective breeding of plants and animals, which is the non-human analogue of eugenics, has, either consciously or by chance, led to new varieties of domesticated plants and animals. A well-established procedure for altering, mostly in desired ways, the genetic constitution of individual members of a given species is to breed only from those whose hereditary properties have desired characteristics. The science of selective breeding is well established as a part of genetics and there is no reason to doubt that it could equally well be practised on human populations, if that were desired. There are many problems to be faced in such eugenically motivated selective breeding programmes. Not the least of these are deciding what to breed for and the implied restriction on the liberty of individuals to reproduce as and when they wish to. In the sense that selective breeding can produce genetic changes, often in rather specific ways, it could be thought of as a form of genetic engineering. The term 'genetic engineering', however, is usually applied only to approaches to genetic manipulation which draw on some of the rather striking possibilities opened up by advances in the molecular understanding of genetics over the past 10 or 20 years, coupled with the ability to manipulate human cells in the laboratory and to grow them in a culture.

The chemistry of the gene

One of the most striking advances in biology since the Second World War has been the discovery of the precise chemical nature of the gene. The gene is now understood to be a molecule of the substance known as deoxyribonucleic acid, or DNA for short. Molecules of DNA can be thought of as very long strings of four different sorts of units, designated by the letters, A, G, C, and T (see Fig. 6.1). It is the order of these units which constitutes the 'language' of the gene and carries the information that is the substance of heredity.

In fact, there are two complementary strings of units intertwined in helical fashion, forming the 'double helix' as first described by Watson and Crick in 1953.

Molecules of DNA have been synthesized in the test-tube, and the

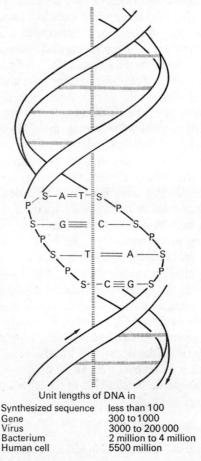

Unit lengths of DNA in
Synthesized sequence   less than 100
Gene   300 to 1000
Virus   3000 to 200 000
Bacterium   2 million to 4 million
Human cell   5500 million

FIG. 6.1. The DNA double helix and the numbers of units in the sequences so far synthesized in the laboratory compared with the numbers present in naturally occurring DNA molecules. Each helical chain is formed of alternate sugar (S) and phosphate (P) groupings; the base pairs (A = T, etc.) link the two.

goal of the modern genetic engineer might be described as the complete synthesis of a set of genes. This would be followed by the incorporation of these genes into a cell in such a way that it would divide and express precisely the properties of the synthesized set of genes. Given that the 'language' of the gene is understood, organisms of desired types could then be synthesized at will. This is a daunting and startling prospect if the knowledge is applied to man. The difficulty, however, lies in the scale of the problem when it is applied to organisms as complicated as man. The DNA sequences that have so far been synthesized are, at most, a few tens of units long, as compared with the total DNA sequence of man, which has some 3000 million units in it. The first step, clearly, is to synthesize one or only a few of the relevant genes. The next step would be to put the gene in its appropriate place, so that it could replace whatever defective gene was there. Apart from the extremely difficult task of synthesizing even just one gene (on average likely to be several hundred to a thousand units long), the problem of getting this to its right place in the genetic material of a cell is formidable.

Experiments on cells in culture

There is also the question of what cells to use. Many sorts of cells can now be grown in culture just as if they were microbes. One of the hopes is to manipulate these in culture and then put them back into the body, having cured whatever defect they carried. As already mentioned, the approach to the problem depends on whether it is wished to cure the disease in the individual or in his offspring. In the former case it is the cells of the diseased tissues or organ that must be changed, while in the latter it is the sperm or the ovum. The sufferer in the latter case is not necessarily cured unless other cells in the body are changed at the same time. Though the techniques of biology may be similar in these two situations, their goals are quite different. Curing the cells which cause the disease—gene therapy— has no further genetic implications. Curing the sperm or ovum cells is, however, a form of eugenics which has implications for the control of future generations. While the prospects of manipulating cells other than the sperm or ova are, at least in a limited number of cases, not too remote, any prospect of curing the sperm or ova, especially on a large scale, is still a long way off. Gene therapy must be clearly distinguished from any manipulation of the genes which has eugenic aims.

The approaches to genetic engineering that have opened up in recent years using cells outside the body growing in a culture are a **natural consequence** of the geneticist's desire to obtain a better understanding of basic cellular mechanisms. This includes the mechanisms which underlie the causes of many diseases, such as cancer. Geneticists are also stimulated by the desire to improve techniques of plant and animal breeding and to create crops and farm animals which yield more in order to cope with the rising world population and the need for adequate nutrition. The origins of this work go back to the British bacteriologist Griffiths, who in trying to understand more about pneumonia in the late 1920s discovered the process known as bacterial transformation. As it later came to be understood, this is a form of 'genetic crossing' in which naked DNA, the chemical substance, is one of the parents and is mixed with whole bacteria as the other parent. Progeny bacteria are obtained which have acquired one or more of their genes from the added DNA.

In the late 1940s, genetic systems for making such crosses were discovered in both bacteria and in the viruses that attacked them. The opportunity that these genetic systems provided to manipulate genetic material of bacteria and their viruses in a controlled way opened up many of the possibilities of molecular biology in the past 20 years. It is, in fact, now possible to contemplate synthesizing the whole, or part, of a naked gene and inserting it back into a bacterium to obtain a desired effect.

Although similar experiments have been tried with human and other mammalian cells in culture with some reported success, these have not, for the most part, been confirmed. Thus, while in principle it might be possible to develop similar systems for human cells, the technical difficulties seem to be almost overwhelming.

Other approaches are, however, available which do not seem to be so remote. One that is now widely used by geneticists and cell biologists depends on the fact that cells of quite different types can be mated together just like bacteria. These matings can be used to make genetic crosses with cells in culture. In this way it is even possible to cross mouse cells with human cells and make a hybrid cell containing both human and mouse genetic material (see Pl. 4(b)). One or more genes can in this way be transferred from a normal cell to an abnormal cell that is to be genetically cured. The difficulty of this approach is that unwanted genetic material may be wrongly transferred. It is also difficult to ensure that the genetic balance of the receiving cell is not upset. Many workers in Britain and elsewhere are now using such systems to do genetics with cells in culture and it has been shown how, in culture, certain genetic 'deficiencies' can be cured in this way and even how a cancer cell may be turned into a normal cell.

Another approach to genetic curing of cells in culture takes its cue again from work with bacteria. This is to use a virus to carry a gene tagged on to its genetic material into the cell which is to be cured. The problem here, once again, is to make sure that the gene gets to its right place in the cell.

The problems of applying these approaches to the genetic curing of cells in culture are still formidable when it comes to curing disease. They are by no means insurmountable but any such approaches to the curing of genetic disease are likely to remain difficult and expensive in the foreseeable future.

Selective breeding

Genetic curing of sperm and ova must, as already mentioned, be considered in quite a different light because of the problems of eugenics and selective breeding they raise. As already discussed in Chapter 4, the problem of eliminating genes from a population is so formidable that it must be questioned whether these most sophisticated techniques of genetics and cell biology can, or should ever, be applied here. While the techniques of genetic engineering may seem more bizarre and frightening than those of selective breeding, the problems they raise for society when applied for eugenic aims are very similar. There is every reason to be as cautious and sceptical about the application of genetic engineering to eugenic ends as there is about the application of selective breeding for the same purpose.

Why, in fact, should the natural methods of reproduction, selection, and mating be tampered with? The answer, in part, is that ever since the beginning of civilization there have been certain constraints, such as the avoidance of incest, on all marriages. (It is interesting to note that chimpanzees also avoid incest.) More recently, the availability of contraceptives, the introduction of safer abortions, and advances in medicine which have prolonged and sometimes ensured the lives of genetically defective people, have meant that the production of future generations is much less subject to the conventional forms of natural selection than it used to be. Radiation released into the atmosphere and chemicals sprayed on the land in recent years have also affected human genes, although probably the amount of change induced is fortunately still slight.

Slowly but surely society is already, if unknowingly, predetermining the future. The problem is whether the process should be directed and accelerated by selective breeding or genetic engineering. To judge this, society has to be reasonably certain that the changes envisaged are for the good of the person directly involved and for the good of society in general. Selective breeding has, as already mentioned, been successfully exploited in plant and animal breeding. There is little doubt that the human race could also be drastically altered by vigorously applying this principle. This option has always been totally unacceptable to the vast majority of people, but it is a fact that selection occurs naturally but slowly in man: indeed this is, after all, how the human species has evolved.

It has for many years been possible to practise selective breeding in humans, and it has also been known that genetic counselling—advising people with genetic diseases of the risks of having affected children—could decrease the incidence of certain diseases. It is not surprising that such counselling has increased in recent years with great effect. As explained in Chapter 4, however, for the majority of severe genetic diseases, which are recessive, counselling has little effect on the frequency of a defective gene in the population; it certainly does not result in a disease being eradicated. But for diseases caused by dominant genes it could eventually decrease the incidence of the disease to that caused by mutations, that is, changes in genes which take place spontaneously.

Cloning

'Cloning' is a word that in recent years has entered the vocabulary of many people. To clone means to produce populations of cells or organisms from a single cell or common ancestor by the normal process of cell division. This bypasses the sexual processes which lead to a scrambling of the genetic material of both parents, and so the products of a clone (except for rare mutations) are genetically identical cells or individuals. Every time a plant is grown from a cutting the plant is, in effect, being cloned. But the application of the idea of cloning to animals is much more difficult. The possibility of doing so follows the much-heralded success of zoologists in producing identical clones of frogs. This has led to the suggestion that the same may be done for humans. If this became possible and the techniques of cloning were coupled with those of genetic engineering of the cells to be cloned, the rather frightening possibility could

Genetic engineering and cloning

emerge of producing armies of identical individuals with carefully planned genetic make-up: a truly Orwellian 1984 spectre. Fortunately 1984 will be with us much sooner that these possibilities, and we can hope that society will be sufficiently forewarned to make sure that the outcome of any such developments is most carefully controlled.

The procedure for cloning a frog is essentially simple. A cell is taken from a frog embryo in an early stage of development and its nucleus injected into an unfertilized egg which has had its cell nucleus removed or killed. In such circumstances with selected species of amphibia, a significant proportion of the transplants develop into adults and the frogs which emerge are similar to the ones that grow from normally conceived embryos but their characteristics correspond to those of the animals from which the nucleus was removed.

In 1971, zoologists at the University of Oxford transferred nuclei from cells grown in the laboratory (rather than from embryos) and succeeded in obtaining a few adult animals from such cell nuclei. These cells were obtained from tadpoles, but no success has yet been reported in the cloning of frogs with cells taken from adult frogs.

Why has it been proved possible to clone amphibia but not, so far, mammals? The zoologists working on cloning think that inherently there is nothing to prevent mammals being cloned, but the techniques necessary to achieve success have not yet been developed. In essence the problem is that mammalian eggs are much smaller and more difficult to manipulate and the conditions of early development in mammals are much more difficult to control. Cloning of humans is still almost as remote a possibility as it has ever been.

Lessons from plant genetics

In the past, many important developments in genetics, particularly concerned with the understanding of the behaviour of chromosomes, have followed work done with plants. Plants offer striking opportunities not available in animals. In several cases it is possible to grow a whole plant from a single cell which has been grown and manipulated in cell culture. This opens up the possibility of manipulating plant cells in culture to produce desired genetic changes of one type or another, and then reconstituting from these manipulated single cells the whole plant. Thus, some of the rather remote possibilities that were discussed above for animals, and for humans in particular, are much more immediate possibilities with plants. The introduction of genetic material from one species of plant into another quite different from it may therefore be envisaged when working with cells in culture. This is out of the question when dealing with crosses in the usual way which, of course, involve the whole organism. Novel combinations of characteristics, many of which could be most important when applied to economic crops, can therefore be envisaged. Here it can indeed be hoped that genetic engineering and its manipulations may solve some of the world's food problems. As an example, new forms of resistance to insect virus and fungal diseases may be introduced. Genes producing more balanced proteins, which are better nutritionally, may be made to play their part. The capacity to fix nitrogen may be introduced into plants that do not have it, making the problem of fertilization much more amenable. All of these developments in plant genetics are to be welcomed because of their important contributions to agriculture.

The future

Whatever the concerns might be about the development of genetic engineering techniques when it comes to their application to man, there should be no concern in their application to plants. Gene therapy, cloning, and the manipulation of genes in ova or sperm are all more of a hope than an expectation in humans at present. But is there any real concern about the way in which research on these topics could develop?

There should be little concern about the future development of gene therapy, for it is a way of curing genetically determined diseases with no effect on future generations. The problem with gene therapy is that it will be a sophisticated method of curing very rare diseases. The result is that when a cure for a particular disease is developed, the cost of each successful diagnosis will be high. Should the money now available therefore be spent on developing the best form of treatment for sufferers of these diseases, or on supporting research based on gene therapy? Questions of this sort bedevil most medical research, and the problems related here are far from unique.

There are greater difficulties associated with cloning and with research based on the manipulation of the genes in sperm and ovum cells, for in principle these can be used to alter man's future. The same

Genetic engineering and cloning

difficulties are evident here that occur when selective breeding programmes are discussed. There must always be substantial reservations about any programme of research designed to alter man's genetic heritage without fully knowing what it is to achieve. There are also the problems of ensuring that any offspring born after the genes of the parent's sperm or ovum are manipulated will be normal. This is a situation analogous to that described in Chapter 3 on artificial fertilization, where one of the concerns of the scientists and doctors involved in the work is that the embryo is not harmed in the manipulation processes.

At the end of 1973 the possible problems of applying the knowledge gained from cloning and genetic engineering did not have to be faced. There is, however, a need for society to be aware of the possible ways in which this research can develop. It is only by being in possession of the full facts that the implications and possible dangers of any line of research can be properly assessed.

7 Social concern and biological advances

THE issues discussed in the previous chapters of this book have several common threads. Not the least of these is that most of the topics covered present dilemmas which society will have to resolve at some time in the near future. The practice of artificial insemination in humans is, for example, long past the stage where a decision is needed on its legality. Similarly the further development of kidney transplantation is being held back because of an Act of Parliament formulated in 1961 when kidney transplantation was no more than an experimental procedure. A time-lag between the application of scientific and medical advances and the adjustments of laws to accommodate them is, of course, necessary but this time should not be allowed to be disproportionately long.

There are also problems associated with how soon society should adopt scientific advances and apply them for the wellbeing of the general population. The Department of Health and Social Security in Britain has in the past shown no reluctance in adopting measures which are clearly of economic benefit to the country. In particular, as mentioned in Chapter 4, there is now in Britain a nation-wide screening programme for the detection of children suffering from phenylketonuria. Similarly, the benefits of both dialysis and kidney transplantation have been appreciated and units are in existence to provide relief for sufferers of kidney diseases which would have proved fatal without treatment. Some of the developments described in previous chapters, which are now emerging from the realms of experiment to that of therapy, do not have such clear-cut economic advantages as the treatment of phenylketonuria and kidney disease.

Social concern and biological advances

For example, the cost of screening all pregnancies in order to detect a mongol child early enough for an abortion to be carried out would be high, and although much suffering would be averted there might not be a direct return to the economy of the money spent on diagnosis. This is in contrast to PKU screening and the prevention of the death of sufferers of kidney diseases.

Although concern has been expressed about the application of some research, as yet there has been no call for research to be controlled from outside the scientific field. Articles have, for example, been written about *in vitro* fertilization which in effect question the wisdom of continuing with the work by pointing out how the knowledge could be misapplied. This concern parallels the way in which artificial insemination of humans received a great deal of publicity in the 1940s and 1950s. The result in the case of artificial insemination was that several inquiries were held to see whether or not the practice should be encouraged, but with no satisfactory or useful outcome. So far, *in vitro* fertilization has not reached the stage where it provides a practicable cure for infertility, and so its large-scale application has not yet been considered. If inquiries and heated discussions, such as surrounded artificial insemination when it made news, are to be avoided, then the implications and wisdom of the use of *in vitro* fertilization should be considered without delay.

Many advances in scientific and medical research are likely to lead to controversy. There is no use denying that developments of the past few years, by which defective foetuses can be identified in time for an abortion to be carried out, will continue to arouse the emotions of many people simply because an abortion is needed to prevent their birth. No one would impose an abortion on any woman and it has been emphasized in this book that a decision to terminate a pregnancy should be solely that of the mother-to-be. In fact, amniocentesis (the procedure of removing a sample of the amniotic fluid which surrounds the foetus) is generally carried out by doctors only when the woman involved has previously stated that she would be willing for the foetus to be aborted if it were affected.

On the other hand, the implications of the development of genetic engineering, in contrast to the other issues discussed, still seem to be a long way in the future. Nonetheless the work now in progress on fusing cells of mice and humans and the cloning of frogs has given ample opportunities for writers with a tendency to be sensational. Cartoonists too have seized the opportunity. The depiction of Mickey

Mouse sitting in a train reading a newspaper and turning to a person sitting next to him and saying, 'Dad, who was Walt Disney?', has certainly appraised many non-scientists, albeit in a sensational way, of this research.

Artificial insemination by donor

Why is there such a gap between the practice of A.I.D. and its legality? The practice has certainly increased since the late 1950s, although the extent of the increase can not readily be estimated because A.I.D. is effectively a clandestine operation for which no official statistics are available.

There is no doubt that, strictly speaking, a child conceived by A.I.D. is illegitimate. Common law, however, has accepted the uncertainties of fatherhood, and a child born during lawful wedlock is now presumed to be legitimate. There is also no doubt that a husband in Britain who registers the birth of an A.I.D. child as his own is committing an offence according to the 1911 Perjury Act. If, however, he leaves blank the space for the father's name on the register of births then the child is illegitimate. This is indeed an unsatisfactory situation. Although it is unlikely that A.I.D. would be raised in the courts in Britain as a ground for divorce, at least if the husband's permission had been obtained in the first place, the fact that it could be makes it all the more necessary to formulate legislation so that A.I.D. can be put on a firm legal footing. This could be done without either encouraging or discouraging the practice so that the emotion which A.I.D. raises in some quarters could be kept to a minimum.

As mentioned in Chapter 2, there are several possible ways of removing the stigma of illegitimacy from an A.I.D. child but they all require significant changes in the law. The register of births could be marked 'A.I.D.' as it is marked at present to show an adoption. Or a law could be passed to make inadmissible as evidence in a court of law that A.I.D. had been carried out. A much more radical suggestion is that the concept of legitimacy be done away with and replaced by a new concept of acceptance or approbation.

Another issue is whether a child born as a result of A.I.D. should, by right, know who his or her biological father is. It seems to be generally accepted that adopted children should be told of their backgrounds, but there has been no widespread discussion of the situation as regards an A.I.D. child. The first question to be asked is whether an A.I.D. child would benefit from knowing his or her background. But whatever the answer to this question is, should it indeed be policy to reveal to everyone the circumstances of his or her birth? This includes not only A.I.D., but also, as mentioned above, adoption and any other factor which makes the birth or conception of a child unusual.

A.I.D. has grown without any direction from government in Britain. This emphasizes the need to keep a watchful eye on the practice of sperm donation. Sperm donors have not, generally speaking, been paid in Britain—in contrast to the United States. If the donor is paid this might perhaps encourage him to conceal facts of his medical history. If fees were paid should sperm donors have to undergo a series of medical tests before their semen was used? And indeed should the donors be tested to see whether they are carriers of some of the more common forms of genetic diseases? These questions are clearly related to whether practitioners of A.I.D. should be registered by the state. The need for registration might soon become more acute as more and more practitioners turn to freezing semen at -196 °C. In the past few years several sperm banks which store semen at these low temperatures have opened in the United States. In principle, the semen so frozen can still be used after storage for many years. The chief use of such banks in the United States has so far been to provide a form of insurance for men about to undergo vasectomies, although it is of course also possible to store donor sperm to be used for A.I.D. With frozen semen available, the practitioner of A.I.D. will not have to rely on a donor providing a sample a few hours before the insemination and so will have a much greater choice of sperm available. This raises the issue of whether there should be any kind of matching of the characteristics of the donor and that of the husband. Indeed, if it is felt that there should be some controls on payment of donors and conditions placed on the storage of semen for long periods of time, then a strong case is made out for the registration of practitioners of A.I.D.

In vitro fertilization

Legal anomalies associated with *in vitro* fertilization work have not arisen because, as yet, it has not been successfully applied to cure infertility. But some of the difficulties which now surround A.I.D. could become associated with *in vitro* fertilization unless it is discussed openly at this stage of its development. The fact that during this work embryos of a few days old are being disposed of will be an insuperable barrier to some people ever agreeing that the work is necessary. However, the use of an intrauterine coil as a contraceptive has a similar effect. The coil does not prevent the sperm fertilizing the ovum; it merely prevents the fertilized ovum attaching itself to the uterine wall and so effectively leads to the abortion of the fertilized ovum.

When the technique of *in vitro* fertilization is restricted to the sperm of the husband and the ovum of his wife, it is effectively a complement to artificial insemination, using the husband's sperm to counteract his infertility, as described in Chapter 2. If, however, sperm or an ovum from a donor who is not the spouse is used, then this is comparable to A.I.D. What restrictions, if any, should be placed on such applications? Should *in vitro* fertilization studies with both sperm and ovum obtained from a married couple be restricted, or should it be welcomed as a way of curing infertility? At present there is little thought of using either a donor ovum or donor sperm in this work, but once the technique is developed such a possibility will certainly arise.

One of the fears which has been expressed is that the manipulation of human fertilized ova outside the body may give rise to birth abnormalities. As described in detail in Chapter 3 experiments with animals have shown no evidence at all of any increase of birth defects above the normal level after manipulation of the fertilized ova. But, of course, there is as yet no such information available on humans.

Fears about *in vitro* fertilization studies have been expressed both in Britain and in the United States. In March 1972, a Panel on Biological Standards and Technological Developments, appointed by the Board of Science and Education of the British Medical Association, made the following statement about the technique.

Co-operation from patients with problems of infertility will allow clinical experimental research to be undertaken which is designed to help with problems of infertility and also present a deeper understanding of the processes of conception. It is most important that such patients be given detailed explanations of the full procedure and implications before any experimentation involving *in vitro* fertilisation is undertaken. An undertaking should be given to use only the husband's spermatozoa in the fertilisation of the ova obtained by laparoscopy, and until more information is available the Panel considers that it would be unethical to use a foster uterus.

It has been envisaged that it may become practicable to diagnose certain foetal abnormalities in fertilised ova at the stage at which they could be implanted in the mother's uterus. If such techniques are perfected without damaging ova, screening for foetal abnormality in this way might be preferable to termination at 16 weeks.

The Panel sees no objection to experimental research so far carried out involving the culture of human fertilised ova, but considers that the implications of this field of research should be kept under review.

At present there is no legislation governing experimental work which involves *in vitro* fertilisation. Whilst it is important to keep the options on development in this field as wide as possible, safeguards governing this work should be established and it is hoped that the Panel on Biological Standards and Technological Developments of the Board of Science and Education will help to provide guide lines for research workers, both medical and non-medical.

In November 1973 an *ad hoc* committee of members of the National Institutes of Health, the National Institute of Mental Health, and the Food and Drug Administration in the U.S.A. published a report on medical experimentation on human subjects. The committee suggested that no research involving transfers of human ova which have been fertilized *in vitro* should take place until animal studies had indicated that it would be safe.

This echoes an editorial in the *Journal of the American Medical Association* in May 1972 which called for a moratorium on further experimentation on the transfer of such early embryos into the uterus.

Looking further into the future, the time will come when embryos will be kept alive for longer *in vitro* than the few days so far shown possible. Efforts are also being made to develop incubators to ensure that premature babies have every chance of life, no matter at what stage of gestation the foetus is removed from its mother's uterus. The question has to be asked whether it will eventually be possible to extend the techniques of *in vitro* fertilization and the development of incubators until they meet, so that a child can be grown without ever having been within a uterus. Some scientists would claim that this is possible given time and money; but is it a desirable goal? Such a development can in fact be thought of as a cure for certain kinds of infertility—for example, where the woman, because she lacks a viable uterus, is unable to have a child. But does the mother perhaps contribute more than oxygen, hormones, and nutrients to her child during gestation? The question, to which there is no ready answer, is what effect has the mother on her child's developing psyche and what would be the long-term consequences of complete development *in vitro*?

Genetic diseases and genetic screening

With the passage of time, more and more diseases are being detected while the foetus is within the uterus. Screening programmes based on these findings are being instituted, and some diseases are gradually being brought under control. Advances in diet have also enabled the effects of some diseases to be alleviated. In general, such medical advances are given every encouragement in the usual form of grants for research. The major problem with the diagnosis of diseases within the uterus is that an abortion is usually the only logical step after a positive diagnosis. There will thus always be some people opposed to this means of preventing the birth of defective children, because of a basic objection to abortion. In a way, the situation is similar to *in vitro* fertilization as described above, where opposition will always be found among those who hold that life is sacred from the moment of fertilization.

Amniocentesis is a technique which will be used increasingly in the future. It is important, however, to realize that there is a limit to its use. This occurs when a disease is so rare that the problems it poses are comparable to the risks involved in the procedure. What the risks are, or indeed will be in the coming years, is difficult to estimate, for with practice the doctor will get better at the task. But it is clear that very rare diseases are not likely to be amenable to diagnosis by amniocentesis.

Is it economically worth while to carry out many amniocenteses in order to prevent the births of severely affected foetuses? Clearly, where affected children die soon after birth, then only a weak case can be made out for early diagnosis. When, on the other hand, affected children are expected to live for a number of years and to be a burden on their parents, family, and society, more serious thought must then be given to preventing their birth. This raises the question of what is a viable foetus. As medical care improves it becomes possible to keep deformed and damaged infants alive for indefinite periods. There has even been a report of an anencephalic child (one born without a brain) being kept alive for several weeks. What should be the guidelines in such cases?

The Board of Science and Education of the British Medical Association in its booklet on professional standards pronounces as follows on the survival of the handicapped:

Social concern and biological advances

Another dilemma raised by medical advances is that in certain instances new treatments enable physically or mentally handicapped patients to survive whilst failing to alleviate the handicap. Such situations are frequently encountered in the fields of paediatrics and geriatrics as well as in the treatment of accident surgery. The decision as to what is best in the interests of the patients must be left to the clinical judgement of the physicians and surgeons.

The onus of making a decision is here fairly and squarely placed on the doctor. This problem is acute in the case of spina bifida children. As mentioned in Chapter 4, there seems to be a glimmer of hope that this disease might soon be predicted from an analysis of the blood of the mother-to-be. At present some 2000 affected children are born annually in Britain. The difficulty with the treatment of this disease is that an operation to close the lesions which such children have is quite likely to leave the child both physically and mentally handicapped. So how is the surgeon to proceed? Should all spina bifida children be operated on, leaving many mentally defective children who are going to live for many years, or should the sufferers receive no treatment at all?

In the past few years criteria have been set for determining whether or not a spina bifida child should be treated. The so-called 'contraindications to active therapy' have been adopted by several paediatricians in Britain, and in 1973 were officially accepted as an allowed practice by the Department of Health and Social Security. Basically the extent of the damage to the child at birth is assessed and, from the knowledge of how such children have fared in the past, it is possible to define a line above which all efforts should be made to save the life of the child and below which no action should be taken. Those who are treated have a reasonable chance of living a life which is not too unpleasant, while those who are not treated generally die within a year of birth, and in no case so far have survived much beyond that.

According to the doctor at the University of Sheffield who pioneered this form of treatment, the nursing staff has accepted the practice. He says that 'there has been no difficulty with the nursing staff, who fully understand the humane purpose and the need for such practice, so long as they are taken into the confidence of the medical staff'. The parents of affected children are given a full explanation of the situation, including the possible therapeutic actions. A prognosis is given of the likely minimal handicap that their child might have if treatment is offered and if everything goes according to plan. A recommendation is then given by the doctor, but the parents are also offered a second opinion.

Such safeguards seem to be reasonable enough. Since the decision not to operate is taken in the expectation that the child will die soon, it may well be asked whether euthanasia (mercy killing) might not be a better solution. It only needs to be said that there is little chance that euthanasia will be legalized, especially in Britain, under any circumstance. Euthanasia is contrary to the general ethos of the practice of medicine, which is directed at saving and prolonging the lives of patients.

Transplantation

It is no exaggeration to say that there is a palpable shortage of kidneys for transplantation in Britain. Many people will continue to die every year until the supply of kidneys increases. The problem is twofold. First, the law which controls the removal of organs from cadavers states that permission to remove the organ must be obtained from the donor before death, or subsequently from the relatives. Time is also of the essence, for kidneys have to be removed within an hour or so of death. Second, there is an understandable reluctance by doctors who have been treating the dead donor to enter into a possibly traumatic interview with distraught relatives to ask them for permission to remove the kidneys. There have been several efforts to increase the supply of kidneys. A committee which sat under the chairmanship of Sir Hector MacLennan in 1969, although divided in the way in which they considered it best to proceed, agreed that a reliable and proved technique, such as kidney transplantation, should not be held back by obsolete laws or by lack of information which might affect the attitude of the medical and nursing staffs of transplant teams.

The nub of the matter is whether Britain should have a 'contracting-in' or a 'contracting-out' system to provide kidneys, or perhaps a combination of both. In December 1972 the Department of Health and Social Security placed its weight behind the present law, which in effect demands a donor either to contract in before death, or for it to be done by the relatives after death. The D.H.S.S. has launched a 'contracting-in' scheme under which potential donors will carry a card signifying that organs may be taken from them in the event of death. It seems unlikely that this will provide a large enough pool of donors, but it should at least provide some publicity for the need for kidneys. Unfortunately, however, there has been little publicity in Britain. A contracting-out scheme would require everyone who objected to his organs being removed in the event of death to register an objection at a central registry. If no objection were registered then it could be assumed that the person agreed to his organs being removed after death. The difficulty with such a system is that the doctor has to make sure within a short time that the name of the dead donor is not registered. In many cases certain identification in a short enough time might not be possible.

An increase in the supply of kidneys for transplantation would release many dialysis machines to cater for people who now receive neither a transplant or dialysis. But it cannot be denied that the removal of organs from corpses will inevitably raise public emotion. It seems that Britain may still have a long way to go before a contracting-out system is accepted similar to those which operate today in Sweden, Israel, Italy, and France.

Genetic engineering

Genetic engineering is still in its infancy so far as its application to the wellbeing of the public is concerned. A great deal of research in this loosely defined field is aimed at curing genetic diseases at the level of the gene itself or of that of the cell. Such research complements other methods, described in Chapter 4, of controlling genetically determined diseases.

Some aspects of genetic engineering, in particular those directed towards altering the sperm and the ovum so that any change will be passed on to future generations, must be closely watched, as indeed must all work aimed at altering the hereditary characteristics of man. This is not to say that the work should be halted or even curtailed, but that wide debate must accompany any application of the results. Such discussions are still some time in the future, as indeed are discussions of the possible effects of the cloning of humans. The research is at such a stage at present that much basic work needs to be done both in cloning and on all forms of genetic engineering before there is any possibility of the results being applied to humans.

The individual and society

An important common aspect of the topics discussed earlier is that the application of many such scientific developments might not always be to the benefit of both the individual and society. Helping infertile couples to have their own children, as discussed in earlier chapters, is a particular example. In the 1970s, with great concern being expressed about overpopulation, it is considered meritorious to avoid having many children and self-indulgent to have them. So how can helping childless couples be justified? Doctors feel that they must do their best for their patients as individuals and that it is not up to them to dissuade childless couples from seeking advice on their infertility. But should governments spend money on alleviating infertility while at the same time they are actively trying to keep the birth-rate down? In 1973, Sir John Peel's panel suggested that artificial insemination by donor be made available through the National Health Service in Britain. If the government were to agree to this it would effectively be spending money on encouraging an increase in the birth-rate, however modest, although the sums involved would be small. In this the advantages to infertile couples would, arguably, outweigh the disadvantages of a minuscule increase in the birth-rate.

Some screening programmes, on the other hand, are of greater benefit to society than to the individual. For example, the screening of teachers for tuberculosis and the screening of airline pilots for colour blindness benefit the children in school and plane passengers more than the teachers and the pilots. Screening newborn children for phenylketonuria aids both the children and their families, but as explained in Chapter 4 the economic advantage to the country of the screening programme seems to be overwhelming. The screening programme for Tay–Sachs disease, also described in Chapter 4, is not specifically supported by public funds in Britain or the United States. The successful identification of affected foetuses in this case is primarily of benefit to the family, because society gains little with such children expected to die within a few years of birth.

One of the most striking recent examples of where a medical technique, if applied universally, would meet individual wishes more than it would benefit society is in the selection of the sex of children. There are now essentially two ways of having a child of a desired sex. First, amniocentesis could be carried out at about the sixteenth week of pregnancy and an analysis of the cells in the amniotic fluid

Social concern and biological advances

would show whether the foetus was male or female. A foetus of the unwanted sex could then be aborted. Second, the X and Y sperm could be separated before an artificial insemination was carried out with the required sperm. At the end of 1973 it was reported that human semen enriched in Y sperm had been obtained in experiments carried out in Germany. No attempted insemination with this enriched sperm was described, but if it were successful then presumably there would be a much greater chance that the foetus would be a boy (although there would be no guarantee, because the semen would only be enriched in Y sperm and not devoid of X sperm).

What if the choice of whether or not to have a child of a given sex were readily available? The first thing to be said is that it would probably reduce the birth-rate, because many couples would decide to have only two children, one of each sex. The problem has, however, been discussed extensively and the following extract from an article by Amitai Etzioni of Columbia University, which appeared in *Science* on 13 September 1969 (vol. 161, pp. 1107–12), points out some of the social consequences which a readily available method of sex selection could bring.

We note, first, that most forms of social behaviour are sex correlated, and hence that changes in sex composition are very likely to affect most aspects of social life. For instance, women read more books, see more plays, and in general consume more culture than men in the contemporary United States. Also, women attend church more often and are typically charged with the moral education of children. Males, by contrast, account for a much higher proportion of crime than females. A significant and cumulative male surplus will thus produce a society with some of the rougher features of a frontier town. And, it should be noted, the diminution of the number of agents of moral education and the increase in the number of criminals would accentuate already existing tendencies which point in these directions, thus magnifying social problems which are already overburdening our society.

Interracial and interclass tensions are likely to be intensified because some groups, lower classes and minorities specifically, seem to be more male oriented than the rest of the society. Hence while the sex imbalance in a society-wide average may be only a few percentage points, that of some group is likely to be much higher. This may produce an especially high boy surplus in lower status groups. These extra boys would seek girls in high status groups (or in some other religious group than their own)—in which they also will be scarce.

On the lighter side, men vote systematically and significantly more Democratic than women; as the Republican party has been losing consistently in the number of supporters over the last generation anyhow, another 5-point loss could undermine the two-party system to a point where Democratic control would be uninterrupted. (It is already the norm, with Republicans having occupied the White House for 8 years over the last 36.) Other forms of imbalance which cannot be predicted are to be expected. 'All social life is affected by the proportions of the sexes. Wherever there exists a considerable predominance of one sex over the other, in point of numbers, there is less prospect of a well-ordered social life. Unbalanced numbers inexorably produce unbalanced behaviour.' (Quoted in J. H. Greenberg, *Numerical sex disproportion: A study in demographic determinism*. University of Colorado Press, Boulder, 1950.)

Society would be very unlikely to collapse even if the sex ratio were to be much more seriously imbalanced than we expect. Societies are surprisingly flexible and adaptive entities. When asked what would be expected to happen if sex control were available on a mass basis, Davis, the well-known demographer, stated that some delay in the age of marriage of the male, some rise in prostitution and in homosexuality, and some increase in the number of males who will never marry are likely to result. Thus, all of the 'costs' that would be generated by sex control will probably not be charged against one societal sector, that is, would not entail only, let us say, a sharp rise in prostitution, but would be distributed among several sectors and would therefore be more readily absorbed.

Clearly, the provision of a choice of the sex of children is not something to be entered into lightly if a reliable and cheap method ever becomes available. It must be asked, however, whether sex choice should be offered in any circumstances. There are many families with several children of the same sex who desire a child of the other sex. Should such a choice be allowed even if the costs are met by the parents? At present this is a much more immediate question to answer than to consider the implications of a widely available method of sex selection.

Where should the money go?

Whether a particular form of treatment benefits society, or individuals, or both is an issue which often confronts decision makers within government. In Britain, the decision on whether or not to implement nation-wide screening programmes based on new medical developments is in the hands of the Department of Health and Social Security. In a society with unlimited funds at its disposal there would presumably be no opposition to implementing programmes which either saved lives or improved the quality of life of sufferers of diseases. Unfortunately, finance is finite, and decisions have to be taken on which programmes to support and which to neglect. Inevitably this leads to controversy. Why, for example, in

Britain is there a national screening programme to detect, soon after birth, children suffering from phenylketonuria while there is no national programme to screen older mothers-to-be for mongol foetuses? The answer lies partly in a cost–benefit analysis which comes out so overwhelmingly on the positive side for screening for PKU that there are no doubts that the benefits of the programme completely outweigh the costs. So far as mongol children are concerned, there is no such overwhelming economic advantage in preventing their birth. Unfortunately, there is no easy way of quantifying the social effects on a family of having a mongol child.

It is perhaps fortunate that developments in scientific and medical research do not always run in parallel. It is therefore highly unlikely that any government will be called upon to decide between increasing finance for better treatment of cystic fibrosis sufferers and to provide the money to attempt to eradicate the disease by a programme of population-wide screening of pregnant women followed by abortion. If, however, a test were developed so that it was known that a foetus would be a cystic fibrosis sufferer and if the government acted in favour of a national screening programme, coupled with abortion of affected foetuses, then the source of subjects to test drugs and improved diet on would dry up. The argument similarly applies for the other inherited diseases in which the sufferers live for many years.

The argument also has more general implications. If it were to prove possible to increase the supply of kidneys for transplantation so that all kidneys before they were transplanted were 'matched' as far as they can be with respect to blood type and other factors, then there would, after some time, be no way of telling how effective the matching was.

A related problem arises in the case of treatment for a previously untreatable disease. If a drug is developed that has just a slight chance of relieving the symptoms of the disease, should it then be prescribed immediately to all sufferers or should it be withheld from some sufferers in order to see by comparison whether the treatment is effective? In this way the patients not on the drug would possibly not be getting the best treatment but this would be the best way of catering for future sufferers of the disease.

But screening programmes do not only compete with other screening programmes for a share of the health budget. Decisions have to be taken on whether to put more money into, for example, accident units, or to support screening programmes. A relevant, though not dominating, question is how many lives will be saved for the expenditure of certain funds in accident units and how does this compare with the return for money spent on screening programmes? And how is an improvement in the quality of life compared with lives saved when proposed programmes are discussed?

On a larger scale it is also important to realize that health programmes compete with other government departments for a slice of the national cake. *Concorde*, the Channel Tunnel, and similar projects are, in a way, competing for funds with proposed programmes to save lives or to improve the quality of life by making use of biomedical advances.

Science, the public, and the press

The interaction between science, medicine, and the press has been greatest in the past few years in the case of transplantation. Kidney and heart transplants have always made news, and in mid-1973 there was also a great deal of publicity given to a bone-marrow transplant to a young child. *In vitro* fertilization studies have also received more than their fair share of publicity in recent years with the press repeatedly inquiring, as indeed is its right, whether the first baby conceived in this way was about to be born.

Similarly the press has always probed to discover the names of both recipients and donors of kidneys and hearts. There is, however, a great deal to be said for preserving anonymity in these cases. The grief of the relatives of a deceased kidney donor is enough of a burden to bear without their also having to put up with calls from the press and the consequent publicity in the media.

Similarly, the few days after a transplant operation are the most worrying for the relatives of the recipient and they usually have little sympathy with the press, which is looking for the 'human aspect' of the story. Knowledge of the publicity might also cause anxiety to the recipient.

There is, however, another side to the story. If a complete barrier of silence were maintained by the hospitals and families about transplant operations, then it would not be known, for example, whether the funds being spent on kidney units were being abused. Without names and personal details it might not be clear that the kidney transplantation advice was being offered, perhaps, just to the rich or some other special segment of society. In this respect, names and stories do much to ensure that there is no bias in the service. The publishing of names also ensures that the current law is adhered to by the surgeons. For whatever case can be put forward to modify the law (see Chapter 5 for details), there is no excuse for abusing the present law. Such a monitoring of medical practice by the press is quite unusual. Most other medical practices, which in their own way are just as helpful either in extending the life of the patient or in making the quality of life better, are not given so much attention by the press. The reason is probably that both heart and kidney transplants arouse emotion because a cadaver is used as a donor.

How can the conflicting needs of the press and the patients and their relatives be met? A case can be made out for names to be released if all the parties concerned, or perhaps only the patient and his family, agree. The hospital authorities could act as the agent through which the press could obtain these names. If such an arrangement existed, it could well be to the benefit of the press, the patient, and the doctor.

There is great unease among the scientific and medical professions that the Press Council lacks the necessary powers to keep journalists within bounds. The Press Council can be contrasted to the General Medical Council, which has very strong powers at its command and is not afraid to use them if necessary. So far as preserving anonymity is concerned, there already is a precedent in the case of juveniles in courts of law and in the case of people 'helping police with their inquiries'. In both these examples there is a legal barrier to publishing the names. No such barrier exists in other examples, such as the case of patients in hospital. If it is thought necessary to restrict the publication of names, must there be legislation or can the National Union of Journalists, for example, impose restrictions on its members which will be binding?

What is important is that the press should present the correct facts and that they should not be distorted for the sake of being sensational. It has to be said that in the past journalists have sometimes fallen down badly in this respect. Scientists and doctors also have not done as much as they should to counteract bad publicity. A particular example concerns a story carried in a Sunday newspaper about *in vitro* fertilization studies which was a complete fabrication according to the doctors named in the story. No complaint was made to the Press Council. The doctor concerned said that this was so because of his previous experience of the Council in dealing with such matters. He also felt that even if an apology were offered by the newspaper, either with or without pressure from the Council, the damage would have been done and the story would have been picked up by the world's press. It would be impossible to have a retraction published in all these papers. To complain to the editor or the Council (the Press Council will not entertain a complaint unless the complainant has first approached the editor and has failed to obtain satisfaction from him) would, furthermore, have generated still more publicity, according to the doctor concerned. It is clear that the best way to counteract bad reporting is for the press to be made fully aware of the facts before they write their stories. This information must come from the scientists and doctors themselves, who must be prepared to make efforts to inform the press.

The need for education

There have been many unfortunate incidents in recent years where the public and the biologist have been at cross-purposes. As mentioned above, these differences have been accentuated in no little way by the press, but perhaps the greatest need now is for non-scientists to be well informed of scientific developments. In this way the public can assess whether or not public money is being abused. That there is a need for the public to be aware of the way in which research is developing is self-evident if the funds come from the public purse. But it is clearly up to the scientist to provide the information for the public, either directly, or indirectly by informing the press.

There is, if anything, an even greater need to keep some particular sections of the community even more closely appraised of the situation. The interactions between science, the law, and religion have been particularly stressed in this book. The lawyers, theologians, and Members of Parliament need to be closely involved with scientists in discussions of the implications of scientific research. In the past, there have been uncomfortably long time-gaps between a scientific discovery, its application, and the consequent readjustment of society. As time progresses and as more and more is written about science, it is imperative that the appropriate people are consulted before the widespread application of scientific developments which affect society. Past experience has shown that without the appropriate exercise of preparing the public, a long-lasting antipathy might result which might damage the chances of society benefiting from the development.

The message for the future is that scientists must be prepared to devote some of their time to educate the public, and that the public must be prepared to look at all aspects of any scientific development and to come to a balanced assessment in full consultation with the scientist.

Suggestions for further reading

THERE is an enormous literature on almost all the topics covered in this short book, and the following list is in no sense comprehensive. The books listed should, however, in turn provide a guide to further reading for those with a deeper interest in the subject.

The list is divided into three sections. The books in the first section deal with the social, ethical, and legal issues raised by biological advances rather than with the scientific questions themselves. A few of these books are collections of papers or discussions from meetings; most of the remainder are highly personal accounts by individual authors. The books in the second list are accounts, written essentially for the non-scientist, of aspects of genetics and biology covered in this book. Those in the third list are for more advanced background reading at university level for those with an academic interest in genetics and related topics.

Social, ethical, and legal issues

Law and the ethics of AID and embryo transfer. CIBA Foundation no. 17. New Series. Elsevier–Excerpta Medica, North Holland (1973).
The second genesis. ALBERT ROSENFELD. Arena Books, New York (1972).
The new genetics and the future of man. Edited by MICHAEL P. HAMILTON. William B. Eerdmans Publishing Company, Grand Rapids, Michigan (1972).
Fabricated man. PAUL RAMSAY. Yale University Press (1970).
Genetics, science and man. *Theological studies*, vol. 33, no. 3, September 1972.
The biocrats. GERALD LEACH. Jonathan Cape, London (1970).
The biological time bomb. GORDON RATTRAY TAYLOR. World Publishing Company, New York, Cleveland (1968).
Genetic fix. AMITAI ETZIONI. Macmillan Publishing Co. Inc. (1973).
The social impact of modern biology. Edited by WATSON FULLER. Routledge and Kegan Paul, London (1971).

Suggestions for further reading

The challenge of life. Roche Anniversary Symposium. *Experienta Supplementum*, **17**. Birkhauser Verlag, Basel and Stuttgart (1972).

Biology and genetics for the layman

The future of man. P. B. MEDAWAR. Mentor Books (1961).
The language of life. GEORGE and MURIEL BEADLE. Anchor Books, Doubleday and Co. Inc., New York (1967).
The theory of evolution. JOHN MAYNARD SMITH. Penguin Books. 2nd edn (1970).
The gift of life. ROY CALNE. Medical and Technical Publishing Co. (1970).

Useful collections of reprints in genetics and biology:
The chemical basis of life. Readings from *Scientific American* with introductions by PHILIP C. HANAWALT and ROBERT H. HAYNES. W. H. Freeman & Co., San Francisco.
Human variation and its origins. A Scientific American reader. Introduction by W. S. LAUGHLIN and R. H. OSBORNE. W. H. Freeman & Co., San Francisco (1967).
Readings in genetics and evolution. A collection of Oxford Biology Readers. Forward by J. J. HEAD. Oxford University Press (1973).

University texts

Basic elementary textbooks
Medical genetics. Edited by VICTOR A. MCKUSICK and ROBERT CLAIBORNE. Hospital Practice Publishing Co., New York (1973).
An introduction to medical genetics. 6th edn. J. A. FRASER ROBERTS. Oxford University Press, London (1973).
Human genetics. VICTOR MCKUSICK. 2nd edn. Prentice Hall and Company (1969).
Genetics in medicine. JAMES S. THOMPSON and MARGARET W. THOMPSON. W. B. Saunders and Co. (1966).
Reproduction in mammals. Book 2: Embryonic and foetal development. Edited by C. R. AUSTIN and R. V. SHORT. Cambridge University Press (1972).

Advanced texts on aspects of biochemical and population genetics

The principles of human biochemical genetics. HARRY HARRIS. North Holland Publishing Co., Amsterdam and London (1971).
The genetics of human populations. L. L. CAVALLI SFORZA and W. F. BODMER. W. H. Freeman & Co., San Francisco (1971).

Index

ABO blood-group system, 19, 75, 84, 91–2
A.I.D., *see* artificial insemination
A.I.H., *see* artificial insemination
abnormalities,
  of animal foetuses, 35–8
  of human foetuses, 39, 42, 43, 45–7, 49–74, 106, 112, 122–4, 129
  see also genetic disease
Abortion Act (1967), 45
abortion, selective, 7, 10, 42, 43, 45, 52, 59, 61–4, 72, 77–82, 111, 117, 122, 127, 129
  cost, 81, 117, 122, 129
  legal aspects, 43, 45
  numbers, in genetic screening programmes, 77–82
  timing, 61
abortion, spontaneous, 31, 39, 47, 64
achondroplasia, 52
adoption, 13, 20, 23–4, 28, 118–19
adrenogenital syndromes, 54, 57
Africa, 2, 72–3
albinism, 57, 68
aminoacidurias, 46, 54–5, 67–8
amniocentesis, 46, 56, 58–64, 77, 81, 117, 122, 126
  cost, 81
  numbers of abnormalities found by, 61–2, 64
  risk to foetus, 60
  timing, 59, 61
amniotic fluid, 46, 53, 58–60, 67, 117, 126
anaesthesia, 34
  local, 59
anencephaly, 58–9, 122

artificial insemination, 4, 10, 116–19
  by donor, 12–29, 39, 42, 43, 118–19
  by husband, 15, 30–1
  matching of donor to parents, 18–20, 119
aspirin, 37
autosomes, 50
azoospermia, 14

Bacteria, 110, 111
BERLIN, Sir ISAIAH, 1
birth defects, *see* abnormalities
birth-rate, 126, 127
blastocyst, 34–6, 38, 44
blindness, 52, 54, 57
blood transfusion, 84
British Medical Association, 13, 41–2, 122

'Carriers', of genetic disease, 50–1, 63–4, 66, 74, 78, 80
cells, 38, 47–8, 106, 109–11, 112–13
  chromosomes in, 38, 46, 50
  clones of, 112–13
  culture of, for diagnosis of abnormalities, 60–2, 63
  development of, 47
  division, 48
  foetal cells in amniotic fluid, 46, 53, 60–1, 63, 126
  genetic content of, 52, 106
  hybrid, 110
chimpanzee, 111
chromosomes, 38–9, 46, 47–50, 60–2, 106, 113
  abnormal number of, 38–9, 46, 60
  defective, 19, 46, 61–2, 106

Chromosomes—*continued*
  in sex cells, 48–9
  sex, 47–50, 61
Church of England, 12, 21–2
circumcision, 50
cloning, 112–13, 125
colour-blindness, 75, 126
congenital disease, *see* genetic disease
contraception, 111
cornea, transplantation of, 83, 93–5, 96
cortisone, 37
counselling, 82, 112
cystic fibrosis, 54, 56–7, 64–7, 77–9, 81, 129
  frequency of occurrence, 64, 77–9

DNA, 107–10
deafness, 52, 54, 56–7
death, definition of, 92
deoxyribonucleic acid, 107–10
Department of Health and Social Security, 7, 69, 87, 100, 116, 123, 124, 128
dialysis treatment, *see* kidney machine
divorce, 12–13, 25, 118
dominant gene, mode of inheritance of, 50–2
Down's syndrome (mongolism), 39, 46–7, 60, 61, 62, 117, 129
Duchenne muscular dystrophy, 49
dwarfism, 52
dystrophia myotonica, 51–2

EDWARDS, R. G., 33–5
egg, *see* ovum
embryo, 2, 5, 30–9, 43, 44, 49, 113, 115, 120–1
  cloning of, 113
  fertilized *in vitro*, 30–9, 120–1
  frog, 113
  frozen, 32, 36–7
  human, 2, 5, 30–1, 43, 44, 115, 120–1
  rabbit, 31–2
  sex of, 49
  sheep, 32
eugenics, 107, 109, 111
euthanasia, 124
eye, *see* corneal transplant

Fallopian tubes, 4, 31, 32–5, 42
FERGUSON-SMITH, M. A., 60, 62
fertilization, 2, 48–9
  in vitro, 5, 30–44, 117, 119–22, 130, 131
Feversham Report (1960) on artificial insemination, 12, 15–16, 20–1, 26, 27, 29
fluoridation, 75
foetus, 34, 35–6, 39, 40
  abnormal, 7, 10, 42, 45–6, 52, 54, 56, 59–63, 74–8, 81, 106, 121–4, 126
'foster' uterus, 5, 32, 41–2, 120
frog, 112–13, 117

Galactosaemia, 54
gene, 8, 19, 46, 49–54, 56, 106–11
  chemistry of, 107–9
  defective, 19, 46–7, 49, 52–7, 78, 81, 106–7, 109, 112
  dominant, 50–2
  linkage of, 50
  recessive, 53–4, 56
  sex-linked, 49–50
gene therapy, 109–11, 114–15
General Medical Council, 131
genetic disease, 5–7, 13, 18–20, 29, 39, 43, 45–82, 106, 112, 114–15, 122–4, 125, 129
  'carriers' of, 50–1, 63–4, 66, 74, 78, 80
  dominant, 50–2, 80
  frequency of occurrence of, 52, 54–5, 57–8
  in artificial-insemination donors, 19, 119
  polygenic, 46, 58–9
  recessive, 50, 53–4, 62
  sex-linked, 49–62
genetic engineering, 3, 8, 41, 106–15, 117, 125
genetic screening, 45, 52–82, 122–4, 126, 129–30
  cost of, 59, 69–70, 78, 81–2, 129–30
  counselling as part of, 82, 112
  efficiency of, 78–9
  for dominant diseases, 52
  for recessive diseases, 54–62
  programmes of, 74–82
German measles, *see* rubella
gonadotrophins, 34

Haemoglobin, 72, 74
haemophilia, 49–51, 53
hamster, embryos of, 32, 37

Index

heart transplantation, 83–4, 92, 104, 130–1
histocompatibility system, 91–2
hormones, balance of, 5, 11, 31, 34–6, 121
   as teratogens, 37
Human Rights, United Nations Universal Declaration of, 11, 20
Human Tissue Act (1961), 94, 95–100, 116
Hunter syndrome, 49
Huntington's chorea, 51–3
Hurler's syndrome, 54
hybrid cell, 110

I.Q., of phenylketonurics, 69, 71
I.U.D., *see* intrauterine device
immune system, 83–5, 88
   suppression in transplant surgery, 91
implantation, 2, 31, 33–8, 43, 44; *see also* transfer
impotence, 14
incest, 111
infant mortality, 45
infectious disease, 5–6, 9
infertility, 4, 11, 12–29, 32–3, 41–2, 120–1, 126
   in Klinefelter's and Turner's syndromes, 47
insulin, 37
intelligence, inheritance of, 4
intrauterine device (I.U.D.), 5, 120
irradiation, 35, 36, 37

Jews, Ashkenazi, 63–4, 74

Kidney machine, 86–7, 105, 116, 125
kidney, transplantation of, 7, 83–4, 85–92, 96–100, 103, 105, 116–17, 124–5, 129–31
   cost of, 88, 105
   success of, 85, 87–90, 105
   supply of donors for, 84–6, 88, 90–2, 96–100, 103, 124–5, 129
Klinefelter's syndrome, 47

Laparoscopy, 33–4, 120
Law Commission, 43
legitimacy, 13, 23–8, 118
Lesch–Nyhan syndrome, 49
life-expectancy, 45

in genetic diseases, 47, 49, 51, 57, 64–5, 70, 72, 78, 81–2
linkage, 50

Malaria, 2, 73, 80
Marfan's syndrome, 52
maturation,
   of ovum, 34, 35, 36
   of sperm, 34
meconium, analysis of for cystic fibrosis, 66
MEDAWAR, Sir PETER, 2
Medical Research Council, 9, 40, 69
meiosis, 48
menstrual cycle, 4, 15, 31, 33
mental defect, severe, 54, 56–7, 68
mental retardation, 46, 49, 51, 123
   in spina bifida children, 58
   in phenylketonuriacs, 68, 71
mitosis, 48
mongolism, 39, 46–7, 60, 61, 62, 117, 129
mouse, 37, 110, 117–18
   embryos of, 32, 33
mucopolysaccharidoses, 54–5
muscular dystrophy, 49
mutagens, 37
mutation, 80, 112

National Health Service, 13, 16, 23, 29, 86–7, 104, 126
neurofibromatosis, 52
nitrogen, fixation of, 114

Oligospermia, 14
organogenesis, 37, 38
ovary, 4–5, 15, 33–5, 42
overpopulation, 126
oviduct, *see* Fallopian tube
ovulation, 15, 36, 37
ovum, 2, 5, 8, 10–11, 14–15, 31–6, 38, 39, 41–2, 114–15, 120–1
   chromosomes in, 38, 48–9, 51
   enucleated, 113
   fertilized *in vitro*, 5, 10–11, 31–6, 39, 41–2, 120–1
   genetic content of, 38, 52–3, 56, 106, 109, 111, 114–15
   maturation of, 34, 35, 36
oxygen, 37, 72, 121

Pancreas, affected by cystic fibrosis, 64–5

Peel Report (1973) on artificial insemination, 13, 16, 18, 20, 23, 26, 29
penicillin, 37
phenylketonuria (PKU), 46, 54, 56, 57, 67–72, 77, 81–2, 116–17, 126, 129
  frequency of occurrence of, 68, 69–70, 77
  metabolic basis of, 67–8
  parenthood for suffers, 71
  screening for, 69–72, 81–2, 126
  treatment of, 69
PKU, *see* phenylketonuria
placenta, 34
plant genetics, 113–14
polygenic disease, 46, 58–9
pregnancy, 5, 11, 20, 35–7, 39, 43, 44, 45, 71, 126
  detection of abnormal foetus in early pregnancy, 43, 45, 52, 54, 57–62, 78–9, 106, 117, 126
  hormone balance in, 5, 11, 36
  in phenylketonuriac mother, 71
  premature baby, 40, 121
Press Council, 131–2

Queen Victoria, 49

Rabbit, 31–2, 36
  embryos of, 32
racial groups, 18–19, 63–4, 72–4
  Ashkenazi Jews, 63–4, 74
  blacks, 72–4
rat, embryos of, 32, 37
recessive disease, *see* genetic disease
recessive gene, mode of inheritance of, 53–4, 56
rhesus antibodies, 75
Roman Catholic Church, 21, 44, 102
rubella, 37, 43

Selective breeding, 107, 111–12
semen, 5, 12, 14–15, 21–3, 119; *see also* sperm
sex, determination of, 48, 126–8
sex ratio, 49
sexual intercourse, 14, 15, 21, 24
sheep, 32, 36
  embryos of, 36
SHETTLES, L. R. B., 34
sickle-cell anaemia, 58, 72–4, 80
skin colour, inheritance of, 19
speech disorders, 51

sperm, 2, 4, 5, 8
  chromosomes in, 38, 48–9, 51
  frozen, 4, 13, 15, 16–18, 28, 29, 32, 119
  genetic content of, 50–3, 56, 106–9, 111, 114–15
  in artificial insemination, 4, 10, 14, 17, 21–2, 24, 26, 29, 31
  in fertilization *in vitro*, 31–4, 39, 41–2, 119–20
  of donor, 10, 14, 17–18, 21–2, 24, 26, 29, 119–20
  of husband, 10, 15, 17, 24, 41–2, 120
  numbers in semen, 14
  selection of, 8, 111
spina bifida, 58–9, 123–4
STEPTOE, P. C., 33–5
sterilization, 29
streptomycin, 37
subfertility, 4, 14–15
syphilis, 75

Tay–Sachs disease, 54, 62–4, 74, 126
teratogenesis, 37–8
'test-tube baby', 3, 30, 39, 40
thalidomide, 37
tissue grafting, *see* transplantation
transfer, of fertilized embryo to uterus, 31–2, 35–6, 121
transplantation, of organs, 7–8, 33, 83–105, 124–5, 129–31
  ethical aspects, 101–5
  legal aspects, 94, 95–100, 124–5
  of cells, 84
  of cornea, 83–4, 93–5, 96
  of fertilized ova, 31–2, 35–6, 39, 121
  of genes, 84
  of heart, 83–4, 92, 104, 130–1
  of kidney, 83–4, 85–92, 96–100, 124–5, 129–31
  of ovary, 33
  rejection of transplant, 83–4, 93
trisomy, 47
  trisomy 21 (mongolism), 46–7
trophoblast, 36
tuberculosis, 75, 126
Turner's syndrome, 47
twins, identical, 83

Ultrasonics, 35, 59
United States of America,
  artificial insemination in, 4, 20, 26
  blood donation in, 103

cost of medical services in, 9, 70, 86, 88
eye banks in, 95
frozen-sperm banks in, 4, 17, 119
heart transplants in, 92
in vitro fertilization in, 43
infertility rate in, 32
phenylketonuria in, 68, 70–1
sickle-cell anaemia in, 73
uterus, 2, 4, 10, 30–5, 39, 40, 41–2, 46, 57, 58, 59, 80, 106, 121–2

'foster' uterus, 5, 32, 41–2, 120

Vaccination, 6, 8–9, 75
vasectomy, 4, 17, 119
viruses, 110, 114
vitamins, 37

World Health Organization, 6

X-rays, 59
  chest, 75

Alan - 3 arrowhead (previous) oes.
put 4 slips in 1 A4 sheet
results table. Check 4 or 5.